# SpringerBriefs in Computer Science

*Series editors*

Stan Zdonik
Peng Ning
Shashi Shekhar
Jonathan Katz
Xindong Wu
Lakhmi C. Jain
David Padua
Xuemin (Sherman) Shen
Borko Furht
V. S. Subrahmanian
Martial Hebert
Katsushi Ikeuchi
Bruno Siciliano

For further volumes:
http://www.springer.com/series/10028

Jiang Wang · Zicheng Liu
Ying Wu

# Human Action Recognition with Depth Cameras

 Springer

Jiang Wang
Northwestern University
Evanston, IL
USA

Ying Wu
Northwestern University
Evanston, IL
USA

Zicheng Liu
Microsoft Research
Redmond, WA
USA

ISSN 2191-5768          ISSN 2191-5776  (electronic)
ISBN 978-3-319-04560-3  ISBN 978-3-319-04561-0  (eBook)
DOI 10.1007/978-3-319-04561-0
Springer Cham Heidelberg New York Dordrecht London

Library of Congress Control Number: 2014930577

Printed on acid-free paper

Springer is part of Springer Science+Business Media (www.springer.com)

# Preface

Action recognition is an enabling technology for many real-world applications, such as human–computer interface, surveillance, video retrieval, senior home monitoring, and robotics. In the past decade, it has drawn a great amount of interests in the research community. Recently, the commoditization of depth sensors has generated much excitement in action recognition from depth sensors. The new depth sensor has enabled many applications that were not feasible before. On one hand, action recognition becomes a lot easier with the depth sensor. On the other hand, people want to recognize more complex actions which present new challenges.

One crucial aspect of action recognition is to extract discriminative features. The depth maps have completely different characteristics from the RGB images. Directly applying features designed for RGB images does not work.

Complex actions usually involve complicated temporal structures, human-object interactions, and person–person contacts. New machine learning algorithms need to be developed to learn these complex structures.

The goal of this book is to bring the readers quickly to the research front in depth sensor-based action recognition, and help the readers to gain deeper understandings of some of the recently developed techniques. We hope this book is useful for both researchers and practitioners who are interested in human action recognition with depth sensors.

This book focuses on the feature representation and machine learning algorithms for action recognition from depth sensors. After presenting a comprehensive overview of the state of the art in action recognition from depth data, the authors provide in-depth descriptions of their recently developed feature representations and machine learning techniques including lower level depth and skeleton features, higher level representations to model the temporal strucrures and human-object interactions, and feature selection techniques for occlusion handling.

## Acknowledgments

We would like to thank our colleagues and collaborators who have contributed to the book: Junsong Yuan, Jan Chorowski, Zhuoyuan Chen, Zhengyou Zhang, Cha Zhang.

I owe great thanks to my wife Ying Xia. This book would not be possible without her consistent support.

Evanston, USA, November 2013                                        Jiang Wang

# Contents

# Chapter 1
# Introduction

**Abstract** Recent years have witnessed great progress in depth sensor technology, which brings huge opportunities for action recognition field. This chapter gives an overview of the recent development of the 3D action recognition approaches, and presents the motivations of the 3D action recognition features, models, and representations in this book.

**Keywords** Action recognition · Depth camera · Depth feature · Human skeleton modeling · Temporal modeling

## 1.1 Introduction

Human has remarkable ability to perceive human actions purely from visual information. We can localize people and objects, track articulated human motions, and analyze human-object interactions to understand what people are doing and even infer their intents. Automatic human action understanding is essential for many artificial intelligence systems, such as video surveillance, human computer interface, sports video analysis, video retrieval, and robotics. For example, to build a human computer interface that intelligently serves people, a system should be able to not only sense the human movement, but also understand their actions and intent.

Despite a lot of progresses, human action recognition is still a challenging task, because human actions are highly articulated, involve human object interactions, and have complicated spatio-temporal structures. Human action recognition systems not only need to extract the low-level appearance and motion information from videos, but also require sophisticated machine learning models to understand the semantic meanings of those information. To achieve this goal, we have to make advances in multiple fronts: *sensory technology* to accurately obtain visual signals from the world, *video/image representation* to describe high-dimensional visual data, *pattern mining* to discover meaningful knowledge, and *machine learning* to learn from large amount of data.

J. Wang et al., *Human Action Recognition with Depth Cameras*,
SpringerBriefs in Computer Science, DOI: 10.1007/978-3-319-04561-0_1,
© The Author(s) 2014

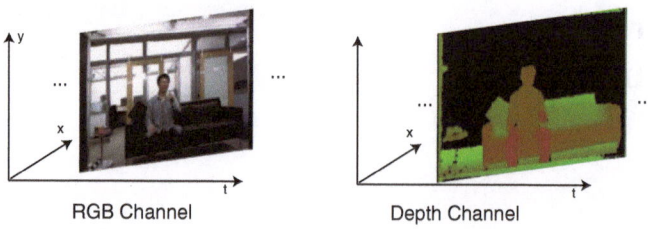

**Fig. 1.1**  Comparison between RGB channels and depth channels

The progress of sensor technologies has led to affordable high-definition depth cameras, such as Microsoft Kinect. Depth camera exploits structural light to sense the depth map in real-time. Pixels in a depth map record the depth of a scene rather than a measure of the intensity of color. The introduction of the depth cameras greatly extends the ability of computer systems to sense the 3D visual world and extract low-level visual information, such as human/object tracking and segmentation. The depth maps are very different from the conventional RGB images, shown in Fig. 1.1. Working in low light levels, and being color and texture invariant, the depth cameras offer several advantages over traditional RGB cameras. However, there are significant flicker noises in depth maps, and the accuracy of depth maps degrades as the distance to the camera increases. Moreover, when only a single depth camera is used, foreground and background occlusions occur frequently, which increases the difficulty of foreground/background segmentation. Novel visual representations and machine learning methods need to be developed in order to fully exploit depth sensors for human action recognition. This presents new scientific challenges that have to be solved to enable next generation applications for security, surveillance, robotics, and human-compute interfaces.

Recent years have witnessed an influx of action recognition methods for depth cameras. The recently developed action recognition methods have made various progresses in feature representation and recognition paradigms. We will give a review of those methods and the recently proposed benchmark 3D action recognition datasets.

## 1.2  Skeleton-Based Features

Kinect provides a powerful skeleton tracking algorithm [1], which outputs 20 3D human joint positions for each frame in real-time. Since the movements of the human skeleton can distinguish many actions, exploiting the Kinect human skeleton output for action recognition is a promising direction.

However, compared to Motion Capture (MoCap) systems [2], which reliably estimate the human joint positions with multiple cameras and joint markers, the human joint positions output by Kinect have lower quality, because Kinect can only use a single depth camera to estimate the human joint positions. The 3D human joint positions output by Kinect are usually noisy when self occlusions or object occlusions occur, or when the camera is looking at the side of the human subject. For example, when one person is crossing his/her arms, bending, or interacting with another person, the skeleton tracker may output very noisy skeleton or even fails. As a result, directly using joint positions does not work well. We need to develop skeleton features that are robust to noise and occlusions.

The tracked skeleton positions are used successfully in a real time dance classification system [3]. This system employs PCA on the torso joint positions to estimate a human torso surface, and represents a human pose with the spherical angles between the limb joint positions and the torso surface. It also employs Fourier transform over time to characterize the temporal structure of actions.

Eigenjoint [4] employs position differences between joints to represent human actions. It computes the position difference of all the pairs of joints within one frame, the joints of two consecutive frames, and the joints of one frame and the initial frame to capture the spatial and temporal configurations of human poses. PCA is applied to the concatenated feature vector to extract "eigenjoints", which are the most important human pose information for action recognition. Finally, a nearest neighbor-based classifier is applied on the eigenjoints features for action classification.

Sequence of most informative joints (SMIJ) [5] finds the top 6 most informative joints according to the variance of joint angle and angular velocity, and construct the feature vectors with the features of these most informative joints. Histogram of 3D joint locations [6] extracts the histogram of spherical coordinates of the joint positions in a coordinates system that uses the hip joint as origin.

Bio-inspired dynamic 3D discriminative Skeleton feature [7] uses linear dynamic systems to model the dynamic medial axis structures of human parts. A discriminative metric is proposed to compare sets of linear dynamics systems for action recognition. The skeleton joints are organized into human parts manually.

The recently developed methods demonstrate that we need to construct robust features and select the most informative joints to achieve robustness against the noise and occlusion in Kinect skeleton tracking systems. However, those methods typically perform joint selection manually or in a unsupervised manner. In this book, we will introduce a novel skeleton representation called *actionlet ensemble*, which not only choose the most discriminative joints in a supervised way but also models the compositional structure of the actions.

## 1.3 Depthmap-Based Features

The skeleton features alone are insufficient to recognize many human actions, because the skeleton features are usually noisy and they fail to capture human-object interactions. For example, the actions "eating" and "drinking" have almost the same human skeleton movement, but they are different because the subject holds different objects in their hands. It is beneficial to construct the features directly from depth maps to better handle human-object interactions.

The simplest way of constructing depthmap-based features is to treat the depth maps as gray images and extract the 2D video features. The widely employed 2D video features include HOG [8], SIFT [9], STIP [10], HOF [10], and kernel descriptors [11]. Those features are shown to work well for 3D object recognition in depth maps in RGB-D object dataset [12].

HOG on Depth Motion Maps (DMM-HOG) [13] applies the HOG descriptor on depth motion maps, which are computed by taking the difference of the depth maps in two consecutive frames, thresholding the difference, and aggregating the difference over time. It extracts three separate DMMs from the front, top, and side views.

Local Depth Pattern (LDP) [14] computes the depth map patterns over local patches. LDP feature uses the interest points as the center of local patches, whose size is scaled inversely by the depth of the interest point. The local patch is divided into a grid. The average depth value is computed for each grid cell, and the difference of the average depth values is computed for every cell pairs. The difference feature vector is used as LDP features.

Although applying 2D video features on depth map sequences achieves reasonably good performance, the depth maps are inherently 3D shapes. We can better characterize the human actions by exploiting the properties of the 3D shapes.

Since one depth map can be treated as a 3D shape, a sequence of depth maps over time forms a 4D spatio-temporal pattern in 4D video volume. Space-time Occupancy Pattern (STOP) [15] roughly characterizes the 4D spatio-temporal patterns of human actions by partitioning the 4D video volume into 4D spatio-temporal cells, and aggregating the occupancy information in each cell. Assuming the person is stationary, the aggregated occupancy information in each cell roughly characterizes the human poses and the human-object interactions.

Histogram of Oriented Normal Vectors (HONV) [16] estimates a 3D normal vector for each point on the 3D shape surface, and constructs a 2D histogram of the normal vectors for each local patch. Histogram of Oriented 4D Normals (HON4D) [17] treats a depth map sequence as a 4D spatio-temporal shape, computes a 4D normal for each point on this shape, and constructs a histogram of the 4D normal vectors. Compared to 3D normal vectors, the 4D normal vectors can capture both the shape and the motion information. The HON4D features also utilize non-uniform 4D quantization by learning the most discriminative non-uniform projectors from training data. The non-uniform 4D quantization is shown to perform much better than uniform 4D quantization.

Depth cuboid [18] extends the cuboid features [19] to depth map sequences. The 3D filtering method is used to remove noise in depth maps before applying the interest point detector. A cuboid similarity feature is extracted from the 3D volume around every depth interest point. The 3D volume is divided into a grid of 3D cells, and the pairwise similarities of the cells are computed as the features. The depth cuboid feature is invariant to spatial and temporal shifts, scales, background clutter, partial occlusions, and multiple motions in the scene.

Researchers also attempted to combine depth maps with RGB videos for action recognition [20] by detecting the interest points in RGB videos, extracting HOG/HOF features and LDP features from RGB videos and depth sequences, respectively, and concatenating the RGB and depth map features. The study shows the information in RGB videos and depth map sequences are complementary to each other.

In this book, we introduce Random Occupancy Pattern (ROP) features, which are obtained through a random sampling and discriminative sparse selection process. The features are robust to partial occlusions thanks to the sparse selection model.

## 1.4 Recognition Paradigms

Building a good feature representation for depth maps is only the first step for human action recognition systems. Human action recognition systems also require effective machine learning algorithms to understand the semantic meanings of human actions. We will give a brief overview of the recognition paradigms for human action recognition using depth cameras.

Direct classification is the simplest classification paradigm. If we can extract robust global feature descriptors for depth map sequences, such as Eigenjoint and SMIJ, we can learn a recognition model by directly applying out-of-the-shelf machine leaning algorithms, such as SVM and Random Forest. Direct classification is very easy to implement, but it is only feasible when the skeleton tracking and depth maps have little noise, and the actions have relatively simple temporal structure.

Bag of words framework is widely used to aggregate local features for recognition systems in RGB videos, it is also successfully employed in action recognition system in depth sequences. Bag of words framework detects a set of interest points from the depth sequences, computes a local descriptor for each interest point, builds a code-book for the local descriptors, and obtains a word histogram vector by quantizing the local descriptors with the codebook. The word histogram vectors are used as feature vectors for classification. Bag of words framework can build action classifiers that are robust to variations in subject location, background, and action speed. However, its performance is largely limited by the discriminative power of local descriptors, and it fails to capture the temporal structure of the actions. When the difference between the actions are determined by the spatio-temporal relationship between the local descriptors, their performance can be sub-optimal.

Nearest neighbor-based method can also be used for action recognition. Finer-earth mover's distance (FEMD) [21] and image-to-class dynamic time warping (I2C-DTW) [22] extend earth mover's distance and dynamic temporal warping to measure the distance between two hand contours, and utilize these distances to build nearest neighbor-based classifiers. Eigenjoints [4] also employs nearest neighbor-based classifiers.

Modeling the temporal structure is essential to successfully modeling complex human actions. Action graph approach [23] models an action as an acyclic directed graph, where every node represents a hidden posture state and every edge represents a transition between the states. Finding the salient posture states is feasible with a clustering algorithm, because it is easier to segment the human subject in depth sequences. The action graph approach can handle temporal alignment, and output the recognition results without having to wait until the action is finished. Maximum Margin Temporal Warping (MMTW) [24] learns phantom action templates via structural latent SVM to represent an action class with maximum discrimination against other classes.

We present two action recognition paradigms in this book. The actionlet ensemble is an effective representation of the spatio-temporal structures of the human actions. The Random Occupancy Pattern (ROP) builds action classifiers that are robust to occlusion via random sampling and discriminative sparse selection.

## 1.5 Datasets

A large number of benchmark datasets have been collected to evaluate depth map action recognition methods. MSR Action3D dataset [25] contains 20 sports actions performed by ten subjects. Each subject performs each action three times. The background of the depth sequences in this dataset is relatively clean, but some actions in this dataset are very similar to each other. This dataset can be used to evaluate the performance of the action recognition algorithms in ideal environments.

MSR Daily Activity3D dataset [26] contains 16 daily activities in the living room, where each subject performs an activity in two different poses: "sitting on sofa" and "standing". When the performer stands close to the sofa or sits on the sofa, the 3D joint positions extracted by the skeleton tracker are very noisy. Moreover, most of the activities involve humans-object interactions. This dataset can be used to evaluate the modeling of human-object interactions and algorithm's robustness to pose changes.

MSR Action Pairs [17] contains 6 pairs of actions, where two actions in a pair involve the interactions with the same object in different ways. For example, the actions "wear a hat" and "take off hat" constitute an action pair. Since the human poses and objects are quite similar for the two actions in each pair, modeling the temporal structure is crucial for successful action recognition on this dataset. This dataset can evaluate the algorithm's ability to model the temporal structure of the actions.

RGBD-HuDaAct dataset [14] captures daily actions for surveillance applications. There are 30 subjects performing these daily activities, which are organized into 14 video capture sessions. In addition, a background activity containing different types of random activities is also captured. This dataset evaluates the benefits of combining RGB and depth videos for action recognition.

MSR Gesture3D dataset [27] is a hand gesture dataset containing a subset of 12 gestures defined by American Sign Language (ASL). All of the gestures used in this experiment are dynamic gestures, where both the shape and the movement of the hands are important for the semantics of the gestures. The self occlusion is more common in this dataset. The hand area is already segmented out using Kinect skeleton tracker. This dataset can be used to evaluate the dynamic hand gesture recognition algorithms in ideal environment.

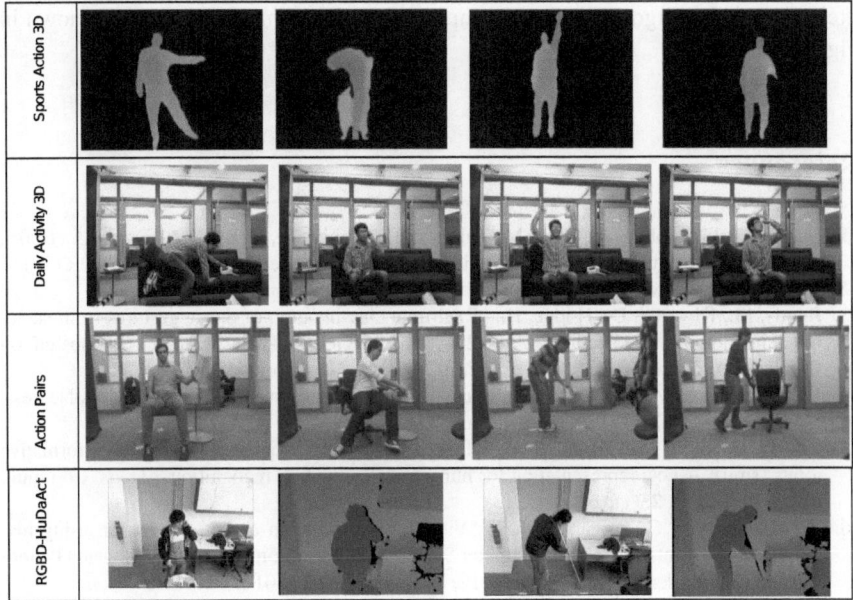

**Fig. 1.2**  Examples of the benchmark action datasets

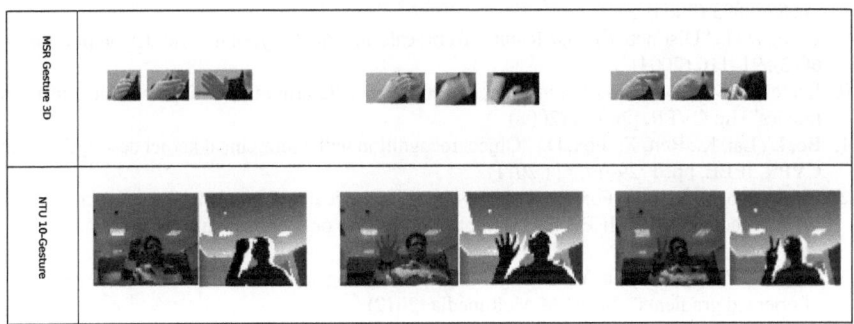

**Fig. 1.3**  Examples of the benchmark gesture datasets

NTU 10-Gesture dataset [21] is collected from 10 subjects, and it contains 10 gestures. Each subject performs 10 different poses for the same gestures. All the gestures in this dataset are static gestures, i.e., their semantics are not determined by the movements of hand. The dataset is challenging because it is collected in uncontrolled environments. Besides, the gestures in this dataset exhibit great variations in hand orientation, scale, and articulation. NTU 10-Gesture dataset evaluates static hand gesture recognition algorithm's real-world performance.

Among these datasets, the MSR Action3D dataset, MSR Daily Activity3D dataset, and MSR Action Pairs dataset contain human skeleton tracked by Kinect human skeleton tracking algorithm. The examples of the benchmark datasets are shown in Figs. 1.2 and 1.3.

# References

1. Shotton, J., Fitzgibbon, A., Cook, M., Sharp, T., Finocchio, M., Moore, R., Kipman, A., Blake, A.: "Real-time human pose recognition in parts from single depth images". In: CVPR (2011)
2. "CMU Graphics Lab Motion Capture Database", http://mocap.cs.cmu.edu/.
3. Raptis, M., Kirovski, D., Hoppe, H.: "Real-time classification of dance gestures from skeleton animation". In: Proceedings of the 2011 ACM SIGGRAPH/Eurographics Symposium on Computer Animation—SCA '11, p. 147. ACM Press, New York, USA (2011)
4. Yang X., Tian, Y.: "EigenJoints-Based Action Recognition Using Naïve-Bayes-Nearest-Neighbor". In: CVPR 2012 HAU3D, Workshop (2012)
5. Ofli, F., Chaudhr, R., Kurillo, G., Vidal, R., Bajcsy, R.: "Sequence of the most informative joints (smij): A new representation for human skeletal action recognition". J. Vis. Commun. Image Represent. **25**(1), 24–38 (2013)
6. Xia, L., Chen, C.C., Aggarwal, J.K.: "View invariant human action recognition using histograms of 3d joints". In: IEEE Computer Society Conference on Computer Vision and Pattern Recognition Workshops (CVPRW), 2012, IEEE, pp. 20–27 (2012)
7. Chaudhry, R., Ofli, F., Kurillo, G., Bajcsy, R., Vidal, R.: "Bio-inspired dynamic 3D discriminative skeletal features for human action recognition". In: HAU3D13 (2013)
8. Dalal, N., Triggs, B.: "Histograms of oriented gradients for human detection". In: CVPR, IEEE, pp. 886–893 (2005)
9. Lowe, D.G.: "Distinctive image features from scale-invariant keypoints". Int. J. Comput. Vision **60**(2), 91–110 (2004)
10. Laptev, I., Marszalek, M., Schmid, C., Rozenfeld, B.: "Learning realistic human actions from movies". In: CVPR, pp. 1–8 (2008)
11. Bo, L., Lai, K., Ren, X., Fox, D.: "Object recognition with hierarchical kernel descriptors". In: CVPR, IEEE, pp. 1729–1736, ( 2011)
12. Lai, K., Bo, L., Ren, X., Fox, D.: "A large-scale hierarchical multi-view RGB-D object dataset". In: Proceedings of the IEEE International Conference on Robotics and Automation (ICRA) (2011)
13. Yang, X., Zhang, C., Tian, Y.: "Recognizing actions using depth motion maps-based histograms of oriented gradients". In: ACM Multimedia (2012)
14. Zhao, Y., Liu, Z., Lu, Y., Cheng, H.: "Combing rgb and depth map features for human activity recognition". In: Signal and Information Processing Association Annual Summit and Conference (APSIPA ASC), 2012 Asia-Pacific, IEEE, pp. 1–4 (2012)
15. Vieira, A.W., Nascimento, E.R., Oliveira, G.L., Liu, Z., Campos, M.M.: "STOP: space-time occupancy patterns for 3D action recognition from depth map sequences". In: 17th Iberoamerican Congress on Pattern Recognition, Buenos Aires (2012)
16. Tang, S., Wang, X., Lv, X., Han, T.X., Keller, J., He, Z., Skubic, M., Lao, S.: "Histogram of oriented normal vectors for object recognition with a depth sensor". In: Computer Vision-ACCV 2012, pp. 525–538. Springer (2013)
17. Oreifej, O., Liu, Z.: "HON4D: histogram of oriented 4D normals for activity recognition from depth sequences". In: CVPR (2013)
18. Xia, L., Aggarwal, J.K.: "Spatio-temporal depth cuboid similarity feature for activity recognition using depth camera". In: CVPR, IEEE, pp. 2834–2841 (2013)
19. Dollár, P., Rabaud, V., Cottrell, G., Belongie, S.: "Behavior recognition via sparse spatiotemporal features". In: 2nd Joint IEEE International Workshop on Visual Surveillance and Performance Evaluation of Tracking and Surveillance, IEEE, pp. 65–72 (2005)

20. Ni, B., Wang, G., Moulin, P.: "Rgbd-hudaact: a color-depth video database for human daily activity recognition". In: Consumer Depth Cameras for Computer Vision, pp. 193–208. Springer (2013)

21. Ren, Z., Yuan, J., Zhang, Z.: "Robust hand gesture recognition based on finger-earth mover's distance with a commodity depth camera". In: Proceedings of the 19th ACM International Conference on Multimedia. ACM, pp. 1093–1096 (2011)

22. Cheng, H., Dai, Z., Liu, Z.: "Image-to-class dynamic time warping for 3d hand gesture recognition". In: IEEE International Conference on Multimedia and Expo (ICME), 2013, IEEE, pp. 1–6 (2013)

23. Li, W., Zhang, Z., Liu, Z.: Expandable data-driven graphical modeling of human actions based on salient postures. IEEE Trans. Circuits Syst. Video Technol. **18**(11), 1499–1510 (2008)

24. Wang, J., Ying, W.: "Learning maximum margin temporal warping for action recognition". In: ICCV, pp. 87–90. ACM (2013)

25. Li, W., Zhang, Z., Liu, Z.: "Action recognition based on a bag of 3d points". In: Human Communicative Behavior Analysis Workshop (In Conjunction with CVPR) (2010)

26. Wang, J., Liu, Z., Wu, Y., Yuan, J.: "Mining actionlet ensemble for action recognition with depth cameras". In: CVPR (2012)

27. Kurakin, A., Zhang, Z., Liu, Z.: "A real-time system for dynamic hand gesture recognition with a depth sensor". In: EUSIPCO (2012)

# Chapter 2
# Learning Actionlet Ensemble for 3D Human Action Recognition

**Abstract** Human action recognition is an important yet challenging task. Human actions usually involve human-object interactions, highly articulated motions, high intra-class variations and complicated temporal structures. The recently developed commodity depth sensors open up new possibilities of dealing with this problem by providing 3D depth data of the scene. This information not only facilitates a rather powerful human motion capturing technique, but also makes it possible to efficiently model human-object interactions and intra-class variations. In this chapter, we propose to characterize the human actions with a novel *actionlet ensemble* model, which represents the interaction of a subset of human joints. The proposed model is robust to noise, invariant to translational and temporal misalignment, and capable of characterizing both the human motion and the human-object interactions. We evaluate the proposed approach on three challenging action recognition datasets captured by Kinect devices, a multiview action recognition dataset captured with Kinect device, and a dataset captured by a motion capture system. The experimental evaluations show that the proposed approach achieves superior performance to the state of the art algorithms.

**Keywords** Actionlet · Frequent item mining · Fourier temporal pyramid · Local occupancy pattern

## 2.1 Introduction

Recognizing human actions has many applications including video surveillance, human computer interfaces, sports video analysis and video retrieval. Despite remarkable research efforts and many encouraging advances in the past decade, accurate recognition of the human actions is still a quite challenging task. There are two major issues for human action recognition. One is the sensory input, and the other is the modeling of human actions that are dynamic, ambiguous and interactive with other objects.

J. Wang et al., *Human Action Recognition with Depth Cameras*,
SpringerBriefs in Computer Science, DOI: 10.1007/978-3-319-04561-0_2,
© The Author(s) 2014

Human motion is articulated in nature. Extracting such highly articulated motion from monocular video sensors is a very difficult task. This difficulty largely limits the performance of video-based human action recognition, as indicated in the studies in the past decade. The recent introduction of the cost-effective depth cameras may change the picture by providing 3D depth data of the scene, which largely eases the task of object segmentation. Moreover, it has facilitated a rather powerful human motion capturing technique [1] that outputs the 3D joint positions of the human skeleton. As we will show in this chapter, although the estimated 3D skeleton alone is not sufficient to solve the human action recognition problem, it greatly alleviates some of the difficulties in developing such a system.

The depth cameras in general produce better quality 3D depth data than those estimated from monocular video sensors. Although depth information alone is very useful for human action recognition, how to effectively combine such 3D sensory data with estimated 3D skeletons is nontrivial. First, the 3D skeleton alone is not sufficient to distinguish the actions that involve human-object interactions. For example, "drinking" and "eating snacks" exhibit very similar skeleton motions. Additional information is needed to distinguish the two actions. Second, human actions may have specific temporal structure. For example, the action "washing a mug" may consist of the following steps: "arriving at the mug", "taking the mug", "arriving at the basin" and "dumping the water". The temporal relationship of these steps is crucial to model such actions. Finally, human actions may have strong intra-class variations. A person may use either his left hand or right hand to make a phone call, and different people have different ways of washing a plate. Modeling these variations is also challenging.

This chapter proposes novel features to represent human actions in depth data. First of all, we propose a new 3D appearance feature called *local occupancy pattern* (LOP). Each LOP feature describes the "depth appearance" in the neighborhood of a 3D joint. Translational invariant and highly discriminative, this new feature is also able to capture the relations between the human body parts and the environmental objects that the person is interacting with. Secondly, to represent the temporal structure of an action, we propose a new temporal representation called *Fourier Temporal Pyramid*. This representation is insensitive to temporal sequence misalignment, robust to noise, and is discriminative for action recognition.

More importantly, we propose a new model called the *Actionlet Ensemble Model*, illustrated in Fig. 2.1. The articulated human body has a large number of kinematic joints, but a particular action may only involve a small subset of them. For example, for right-handed people, action "drinking" typically involves joints "right wrist" and "head". Thus the combinational feature of the two joints is a discriminative feature. For left-handed people, action "drinking" typically involves joints "left wrist" and "head". Therefore the combinational feature of joints "left wrist" and "head" is another discriminative feature for this action. Therefore, we introduce the concept of *actionlet*. An *actionlet* is a conjunction of the features for a subset of the joints. As the number of possible *actionlets* is enormous, we propose a novel data mining solution to discover *discriminative actionlets*. An action is then represented as an *actionlet ensemble*, which is a linear combination of the *actionlets* whose discriminative

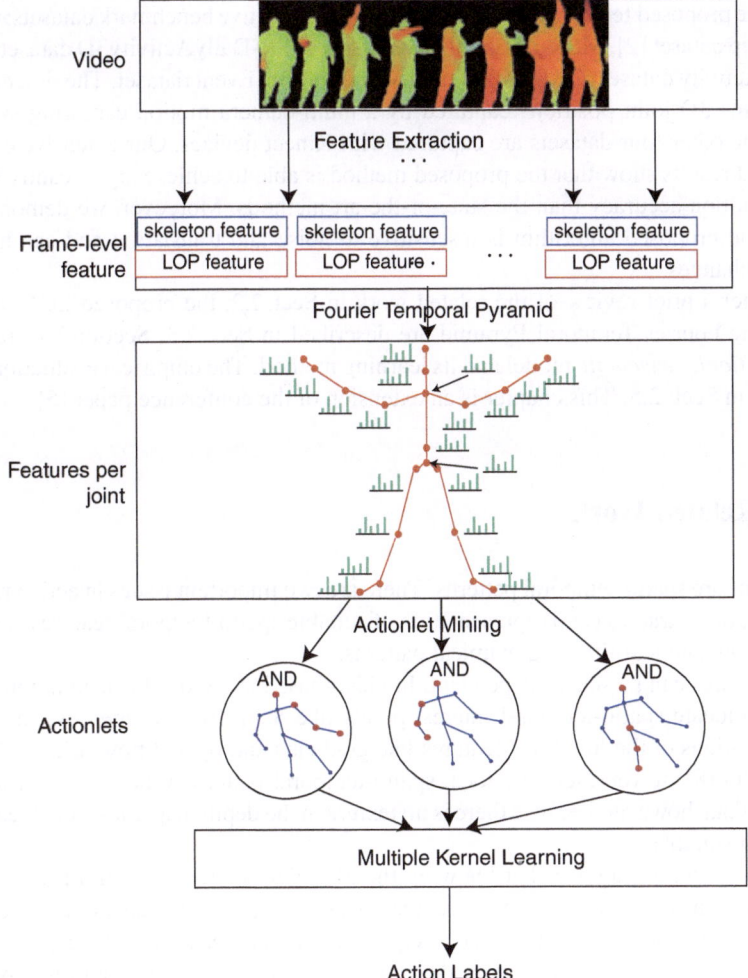

Video

Feature Extraction
···

Frame-level
feature

| skeleton feature | skeleton feature | | skeleton feature |
| LOP feature | LOP feature · | ··· | LOP feature |

Fourier Temporal Pyramid

Features per
joint

Actionlet Mining

AND    AND    AND

Actionlets

Multiple Kernel Learning

Action Labels

**Fig. 2.1** The general framework of the proposed approach

weights are learnt via a multiple kernel learning method. This new action model is more robust to the errors in the features, and it can better characterize the intra-class variations in the actions.

Our main contributions include the following three aspects. First, this chapter proposes the *actionlet ensemble* model as a new way of characterizing human actions. Second, we propose a novel feature called local occupancy pattern, which is shown through our extensive experiments to be well suitable for the depth data-based action recognition task. Third, the proposed Fourier temporal pyramid is a new representation of temporal patterns, and it is shown to be robust to temporal misalignment and noise.

The proposed features and models are evaluated on five benchmark datasets: CMU MoCap dataset [2], MSR-Action3D dataset [3], MSR-DailyActivity3D dataset, Cornell Activity dataset (CAD-60) [4] and Multiview 3D Event dataset. The first dataset contains 3D joint positions captured by a multi-camera motion capturing system, and the other four datasets are captured with Kinect devices. Our extensive experimental results show that the proposed method is able to achieve significantly better recognition accuracy than the state-of-the-art methods. Moreover, we demonstrate that the proposed algorithm is insensitive to noise and translation and can handle view changes.

After a brief review of the related work in Sect. 2.2, the proposed LOP feature and the Fourier Temporal Pyramid are described in Sect. 2.3. Section 2.4 presents the *actionlet ensemble* model and its learning method. The empirical evaluations are given in Sect. 2.5. This chapter is an extension of the conference paper [5].

## 2.2 Related Work

Actions are spatio-temporal patterns. There are two important issues in action recognition: the extraction and representation of suitable spatio-temporal features, and the modeling and learning of dynamical patterns.

Features can be sensor-dependent. In video-based methods, it is a common practice to locate spatio-temporal interest points like STIP [6], and then use the local distributions of the low-level features like gradients and optical flow (e.g., HOF [7] or HOG [8]) to represent the local spatio-temporal pattern. When we want to use depth data, however, because there is no texture in the depth map, these local features are not suitable.

It is generally agreed that knowing the 3D joint positions is helpful for action recognition. Multi-camera motion capture (MoCap) systems [9] can produce accurate 3D joint positions, but such special equipment is marker-based and expensive. It is still a challenging problem to develop a marker-free motion capturing system using regular video sensors. Cost-effective depth cameras have been used for motion capturing, and produced reasonable results, despite the noise when occlusion occurs. Because of the difference in the motion data quality, the action recognition methods designed for MoCap data might not be suitable for depth cameras.

In the literature, there have been many different temporal models for human action recognition. One way to model the human actions is to employ generative models, such as a Hidden Markov model (HMM) and Conditional Random Field (CRF). Lv and Nevatia [10] used HMM over pre-defined relative positions obtained from the 3D joints. Han et al. [11] used CRF over 3D joint positions. Similar approaches are also proposed to model human actions in normal videos [12, 13]. The 3D joint positions that are obtained via skeleton tracking from depth maps sequences are generally more noisy than the MoCap data. When the difference between the actions is small, without careful selection of the features, determining the accurate states is usually difficult, which undermines the performance of such generative models. Moreover, with limited amount of training data, training a complex generative model is prone to overfitting.

Temporal patterns can also be modeled by a linear dynamical systems or a non-linear Recurrent Neural Network [14]. Although these approaches are good models for time series data and are robust to temporal misalignment, it is generally difficult to learn these models from limited amount of training data.

Another method for modeling actions is the dynamic temporal warping (DTW) [15], which defines the distance of two time series as their edit distance. The action recognition can be done through nearest-neighbor classification. DTW's performance heavily depends on a good metric to measure the frame similarity. Moreover, for periodic actions (such as "waving"), DTW is likely to suffer from large temporal misalignment thus degrading classification performance [16].

Different from these approaches, we propose a *Fourier Temporal Pyramid* for temporal pattern representation. The Fourier temporal pyramid is a descriptive model. It does not require complicated learning as in the generative models (e.g., HMM, CRF and dynamical systems), and it is much more robust than DTW to noise and temporal misalignment.

In the actions with a complex articulated structure, the motions of the individual parts may be correlated. The relationship among these parts (or high-order features) is often more discriminative than the individual ones. Such combinatorial features can be represented by stochastic AND/OR structures. This idea has been pursued for face detection [17], human body parsing [18], object recognition [19], and human object interaction recognition [20]. This chapter presents an initial attempt of using the AND/OR ensemble approach for action recognition. We propose a novel data mining solution to discover the discriminative conjunction rules based on [21], which is a branch-and-bound algorithm that guarantees to find all the frequent patterns efficiently, and apply multiple kernel learning framework to learn the ensemble. Other work that involves learning the interactions of human joints include poselet model [22] and phraselet model [23]. Poselet has been successfully applied in action recognition by mining discriminative appearance patterns to classify actions [24]. These models learn the relationship among human parts in annotated images.

Recently, a lot of efforts have been made to develop features for action recognition in depth data and skeletons. Li et al. [3] represents each depth frame as a bag of 3D points along the human silhouette, and utilizes HMM to model the temporal dynamics. Wang et al. [25] learns semi-local features automatically from the data with an efficient random sampling approach. Vieira et al. [26] also uses spatio-temporal occupancy patterns, but all the cells in the grid have the same size, and the number of cells is empirically set. Yang and Tian [27] proposes a dimension-reduced skeleton feature, and [28] develops a histogram of gradient feature over depth motion maps. Ofli et al. [29] selects most informative joints based on the discriminative measures of each joint. Yun et al. [30] utilizes distances between all pairs of joints as features and multiple instance learning for feature selection. Raptis et al. [31] utilize Kinect cameras to recognizes dance actions. Chaudhry et al. [32] uses linear dynamic systems to model the dynamic medial axis structures of human parts and proposes discriminative metrics for comparing sets of linear dynamics systems for action recognition, but it organizes skeleton joints into human parts manually rather than automatically learns from data. Our work is the first attempt to model the structure and relationship among the human parts and achieves state-of-the-art performance on multiple benchmark datasets.

## 2.3  Spatio-Temporal Features

This section gives a detailed description of two types of features that we utilize to represent the actions: the 3D joint position feature and the local occupancy pattern (LOP). These features can characterize the human motions as well as the interactions between the objects and the human. In addition, the Fourier Temporal Pyramid is proposed to represent the temporal dynamics. The proposed features are invariant to the translation of the human body and robust to noise and temporal misalignment. The orientation normalization method, which can improve the proposed method's robustness to human orientation changes, is also discussed.

### 2.3.1  Invariant Features for 3D Joint Positions

The 3D joint positions are employed to characterize the motion of the human body. One key observation is that representing the human movement as the pairwise relative positions of the joints results in more discriminative features.

For a human subject, 21 joint positions are tracked by the skeleton tracker [1] and each joint $i$ has 3 coordinates $p_i(t) = (x_i(t), y_i(t), z_i(t))$ at a frame $t$. The illustration of the skeleton joints are shown in Fig. 2.2. The coordinates are normalized so that the motion is invariant to the initial body orientation and the body size. The details of the orientation normalization can be found in Sect. 2.3.4.

For each joint $i$, we extract the pairwise relative position features by taking the difference between the position of joint $i$ and any other joint $j$:

$$p_{ij} = p_i - p_j, \qquad (2.1)$$

The 3D joint feature for joint $i$ is defined as:

$$p_i = \{p_{ij}|i \neq j\}.$$

Although enumerating all the joint pairs introduces some information that may be irrelevant to our classification task, our system is capable of selecting the joints that are most relevant to our recognition task. The selection will be handled by the *actionlet* mining algorithm as discussed in Sect. 2.4.

Relative joint position is actually a quite intuitive way to represent human motions. Consider, for example, the action "waving". It can be interpreted as "arms above the

**Fig. 2.2** The human joints tracked with the skeleton tracker [1]

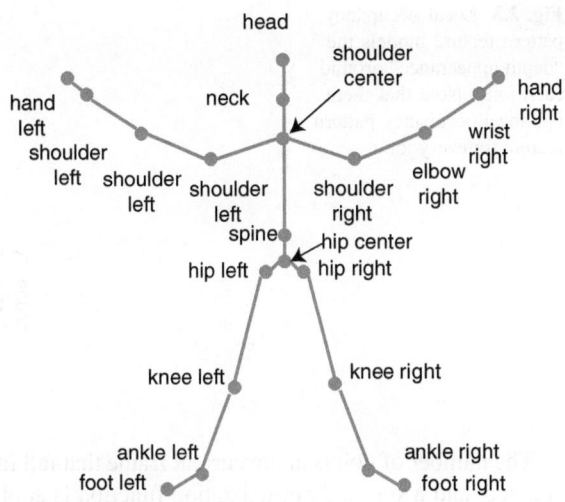

shoulder and moving left and right". This can be effectively characterized through the pairwise relative positions.

### 2.3.2 Local Occupancy Patterns

Using the 3D joint positions alone is insufficient to represent an action, especially when an action includes the interactions between human subject and other objects. Therefore, it is necessary to design a feature to describe the local "depth appearance" for the joints. In this chapter, the interaction between the human subject and the environmental objects is characterized by the *Local Occupancy Patterns* or LOP at each joint. For example, suppose a person is drinking a cup of water. When the person fetches the cup, the space around his/her hand is occupied by the cup. Afterwards, when the person lifts the cup to his/her mouth, the space around both the hand and the head is occupied. The occupancy information can be useful to characterize this interaction and to differentiate the drinking action from other actions.

In each frame, as described below, an LOP feature computes the local occupancy information based on the 3D point cloud around a particular joint. The temporal dynamics of these occupancy patterns can discriminate different types of interactions. An illustration of the spatial-temporal occupancy pattern is shown in Fig. 2.3. Note that we only draw the LOP box for a single joint in Fig. 2.3, but in fact, a local occupancy pattern is computed for every joint.

At frame $t$, we have the point cloud generated from the depth map of this frame. For each joint $j$, its local region is partitioned into $N_x \times N_y \times N_z$ spatial grid. Each bin of the grid is of size $(S_x, S_y, S_z)$ pixels. For example, if $(N_x, N_y, N_z) = (12, 12, 4)$ and $(S_x, S_y, S_z) = (6, 6, 80)$, the local $(72, 72, 320)$ region around a joint is partitioned into $12 \times 12 \times 4$ bins, and the size of each bin is $(6, 6, 80)$.

**Fig. 2.3** Local occupancy
pattern feature models the
"depth appearance" around
each joint. Note that there
is a local occupancy pattern
feature for every joint

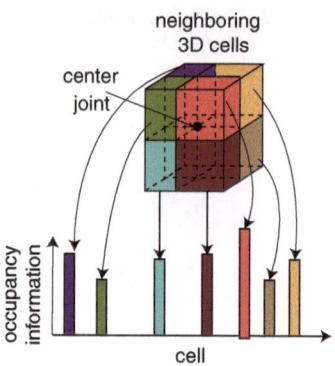

The number of points at the current frame that fall into each bin $b_{xyz}$ of the grid is
counted, and a sigmoid normalization function is applied to obtain the feature $o_{xyz}$
for this bin. In this way, the local occupancy information of this bin is:

$$o_{xyz} = \delta( \sum_{q \in \text{bin}_{xyz}} I_q) \tag{2.2}$$

where $I_q = 1$ if the point cloud has a point in the location $q$ and $I_q = 0$ otherwise.
$\delta(.)$ is a sigmoid normalization function: $\delta(x) = \frac{1}{1+e^{-\beta x}}$. The LOP feature of a joint
$i$ is a vector consisting of the feature $o_{xyz}$ of all the bins in the spatial grid around
the joint, denoted by $o_i$.

### 2.3.3 Fourier Temporal Pyramid

Two types of features are extracted from each frame $t$: the 3D joint position features
$p_j[t]$, and the LOP features $o_j[t]$. In this subsection, we propose the Fourier Temporal
Pyramid to represent the temporal patterns of these frame-level features.

When using the current cost-effective depth camera, we always experience noisy
depth data and unreliable skeletons. Moreover, temporal misalignment is inevitable.
We aim to design a temporal representation that is robust to both noisy data and the
temporal misalignment. We also want such temporal features to be a good representation of the temporal structure of the actions. For example, one action may contain
two consecutive sub-actions: "bend the body" and "pick up".

The proposed Fourier Temporal Pyramid is a descriptive representation that sat-
isfies these properties. It is partly inspired by the Spatial Pyramid approach [33].
In order to capture the temporal structure of the actions, in addition to the global
Fourier coefficients, we recursively partition the action into a pyramid, and use the
short time Fourier transform for all the segments, as illustrated in Fig. 2.4. The final
feature is the concatenation of the Fourier coefficients from all the segments.

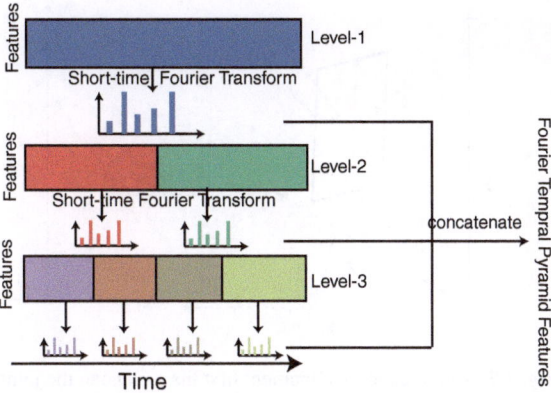

**Fig. 2.4** A illustration of the Fourier Temporal Pyramid

For each joint $j$, let $g_j = (p_j, o_j)$ denote its overall feature vector, where $p_j$ is its 3D pairwise position vector and $o_j$ is its LOP vector. Let $N_j$ denote the dimension of $g_j$, i.e., $g_j = (g_1, \ldots, g_{N_j})$. Note that each element $g_n$ is a function of time and we can write it as $g_n[t]$. For each time segment at each pyramid level, we apply Short Fourier Transform [34] to the element $g_n[t]$ and obtain its Fourier coefficients, and we utilize its low-frequency coefficients as features. The Fourier Temporal Pyramid feature at joint $j$ is defined as the low-frequency coefficients at all levels of the pyramid, and is denoted as $G_j$.

The proposed Fourier Temporal Pyramid feature has several benefits. First, by discarding the high-frequency Fourier coefficients, the proposed feature is robust to noise. Second, this feature is insensitive to temporal misalignment, because a temporally translated time series has the same Fourier coefficient magnitudes. Finally, the temporal structure of the actions is characterized by the pyramid structure.

### 2.3.4 Orientation Normalization

The proposed joint position features and local occupancy patterns are generally not invariant to the human orientation. In order to make the system more robust to human orientation changes, we perform the orientation normalization using the tracked skeleton positions. An illustration of the orientation normalization procedure is shown in Fig. 2.5.

In our experiment, we employ the up-right pose for orientation normalization. We find the frames where the human is approximately in an up-right pose, and use the pose of these frames for orientation alignment. If there is no up-right pose in a sequence, we do not perform orientation normalization for this sequence. For each frame where the human subject is in an up-right pose, we fit a plane to the joints "head", "neck", "hip", "left shoulder", and "right shoulder". The plane normal is used for orientation normalization.

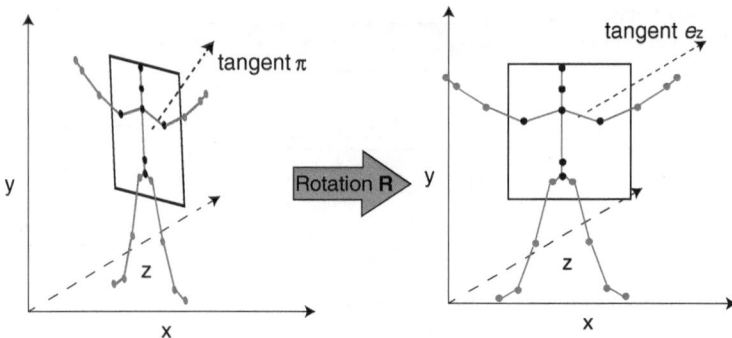

**Fig. 2.5** The orientation alignment first fits a plane to the joints shown as *red* in the figure. Then, we compute a rotation matrix that rotates this plane to the x-y plane

Denote the 3D positions of the joints "head", "neck", "hip", "left shoulder", and "right shoulder" by $p_1, p_2, \cdots, p_5$, respectively. The plane $f(p) = \pi^T[p; 1] = 0, \|\pi\|^2 = 1$, that best fits these joints can be found by minimizing the sum of the distances of the points $p_1, p_2, \cdots, p_5$ to the plane:

$$\min_\pi \sum_{i=1}^{5} \|f(p_i)\|^2 = \min_\pi \|P\pi\|^2 \qquad (2.3)$$
$$s.t. \|\pi\|^2 = 1$$

where $P$ is an constraint matrix defined as

$$\begin{bmatrix} p_1 & p_2 & p_3 & p_4 \\ 1 & 1 & 1 & 1 \end{bmatrix}^T \qquad (2.4)$$

The plane parameters $\pi = [\pi_x; \pi_y; \pi_z; \pi_t]$ that minimize Eq. (2.3) are the right singular vector of $P$ corresponding to the smallest singular value, which can be found by singular value decomposition.

In addition, we employ RANSAC procedure [35] to robustly estimate the plane. The RANSAC procedure iterates between the plane fitting step and the outlier detection step. The plane fitting step fits a plane to the non-outlier points by solving Eq. (2.3). The outlier detection step identifies the points that are too far from the plane as the outliers. The RANSAC procedure is more robust to the outliers of the 3D joint positions. When some joints are incorrectly tracked or the human pose we employ is not precisely upright, the RANSAC procedure can still robustly find the correct plane with small error.

To use the fitted plane for orientation normalization, we find a rotation matrix $R$ that maps orientation of the plane $f(p) = \pi^T[p; 1] = 0$ to the x-y plane: $u(p) = e_z[p; 1] = 0$, where $e_z$ is the vector $[0; 0; 1; 0]$. Denote the normal of the plane $f(p) = 0$ and $u(p) = 0$ as

$$\pi' = \frac{[\pi_x; \pi_y; \pi_z]}{\|[\pi_x; \pi_y; \pi_z]\|_2} \tag{2.5}$$

$$e'_z = [0; 0; 1] \tag{2.6}$$

This is equivalent as rotating the plane normal from $\pi'$ to $e'_z$, shown in Fig. 2.5. The rotation axis $x$ and rotation angle $\theta$ of the rotation matrix $R$ can be found as:

$$x = [x_1; x_2; x_3] = \frac{\pi' \times e'_z}{\|\pi' \times e'_z\|} \tag{2.7}$$

$$\theta = \cos^{-1}\left(\frac{\pi'.e'_z}{\|\pi'\|.\|e'_z\|}\right) \tag{2.8}$$

Then the rotation matrix $R$ can be defined according to exponential map:

$$R = I \cos\theta + A \sin\theta + (1 - \cos\theta)xx^T \tag{2.9}$$

where $A$ is a skew-symmetric matrix corresponding to $x$

$$A = \begin{bmatrix} 0 & -x_3 & x_2 \\ x_3 & 0 & -x_1 \\ -x_2 & x_1 & 0 \end{bmatrix} \tag{2.10}$$

When there are more than one frame with up-right pose, orientation normalization utilizes the average of the fitted plane normals of all the up-right poses in this sequence.

This rotation matrix can be applied to the 3D joint positions and 3D point cloud of all the frames for orientation normalization.

In addition to orientation normalization, we also perform scale normalization. The scale of the body can be estimated from the average pairwise distances of the skeleton joints "head", "neck", "hip", "left shoulder", and "right shoulder".

## 2.4 Actionlet Ensemble

To deal with the errors of the skeleton tracking and better characterize the intra-class variations, an *actionlet ensemble* approach is proposed in this section as a new representation of human actions.

An *actionlet* is defined as a conjunctive (AND) structure on the base features. One base feature is defined as the Fourier Temporal Pyramid features of an individual joint. A novel data mining algorithm is proposed to discover the *discriminative actionlets*, which are highly representative of one action and highly discriminative compared to the other actions.

Once we have mined a set of discriminative actionlets, a multiple kernel learning [36] approach is employed to learn an actionlet ensemble structure that combines these discriminative actionlets.

### 2.4.1 Mining Discriminative Actionlets

The human body consists of a large number of kinematic joints, but a particular action may only involve a small subset of them. For example, for right-handed people, action "calling cellphone" typically involves joints "right wrist" and "head". Therefore, the combinatorial feature of the two joints is a discriminative feature. Moreover, strong intra-class variation exists in some human actions. For left-handed people, action "calling cellphone" typically involves joints "left wrist" and "head". Therefore, the combinatorial feature of joint left wrist and head is another discriminative feature for this action. We propose the *actionlet ensemble* model to effectively characterize the combinatorial structure of human actions. An *actionlet* is a conjunction (AND) of the features for a subset of the joints. We denote an *actionlet* as its corresponding subset of joints $S \subseteq \{1, 2, \cdots, N\}$, where $N$ is the total number of joints. Since one human action contains an exponential number of the possible actionlets, it is time consuming to construct an ensemble from all of the possible actionlets. In this section, We propose an effective data mining technique to discover the *discriminative actionlets*.

We employ the training data to determine whether an actionlet is discriminative. Suppose we have the training pairs $(x^{(i)}, y^{(i)})$, where $x^{(i)}$ is the features of $i$-th example and $y^{(i)}$ is the label of the $i$-th example. In order to determine how discriminative each individual joint is, a SVM model is trained on the feature $G_j$ of each joint $j$. For each training example $x^{(i)}$ and the SVM model on the joint $j$, the probability that its classification label $y^{(i)}$ is equal to an action class $c$ is denoted as $P_j(y^{(i)} = c|x^{(i)})$, which can be estimated from the pairwise probabilities by using pairwise coupling approach [37].

Since an actionlet takes a conjunctive operation, it predicts $y^{(i)} = c$ if and only if every joint $j \in S$ (the joint contained in this actionlet) predicts $y^{(i)} = c$. Thus, assuming the joints are independent, the probability that the predicted label $y^{(i)}$ is equal to an action class $c$ given an example $x^{(i)}$ for an actionlet $S$ can be computed as:

$$P_S(y^{(i)} = c|x^{(i)}) = \prod_{j \in S} P_j(y^{(i)} = c|x^{(i)}) \tag{2.11}$$

Define $\mathcal{X}_c$ as the set of the training data with class label $c$: $\{i : t^{(i)} = c\}$. For a discriminative actionlet, the probability $P_S(y^{(i)} = c|x^{(i)})$ should be large for some data in $\mathcal{X}_c$, and be small for all the data that does not belong to $\mathcal{X}_c$. Define the confidence score for actionlet $S$ as

$$\text{Conf}_S = \max_{i \in \mathcal{X}_c} \log P_S(y^{(i)} = c|x^{(i)}) \tag{2.12}$$

and the ambiguity score for actionlet $S$ as

$$\text{Amb}_S = \frac{\sum_{i \notin \mathcal{X}_c} \log P_S(y^{(i)} = c | x^{(i)})}{\sum_{i \notin \mathcal{X}_c} 1} \tag{2.13}$$

The discriminativeness of an actionlet $S$ can be characterized by its confidence score $\text{Conf}_S$ and ambiguity score $\text{Amb}_S$. A discriminative actionlet should exhibit large confidence score $\text{Conf}_S$ and small ambiguity score $\text{Amb}_S$. Since one action contains an exponential number of actionlets, it is time consuming to enumerate all actionlets. We propose an Aprior-based data mining algorithm that can effectively discover the discriminative actionlets.

An actionlet $S$ is called an $l$-actionlet if its cardinality $|S| = l$. One important property of the actionlet is that if we add a joint $j \notin S$ to an $(l - 1)$-actionlet $S$ to generate an $l$-actionlet $S \cup \{j\}$, we have $\text{Conf}_{S \cup \{j\}} \leq \text{Conf}_S$, i.e., adding a new joint into one actionlet will always reduce the confidence score.

As a result, the Aprior mining process [21] can be applied to select the actionlets with large $\text{Conf}_S$ and small $\text{Amb}_S$. The Aprior-based algorithm is essentially a branch and bound algorithm that effectively prunes the search space by eliminating the actionlets that do not have the confidence score larger than the threshold. If the confidence score $\text{Conf}_S$ of an actionlet $S$ is already less than the confidence threshold, we do not need to consider any actionlets $S'$ with $S' \supset S$, because the confidence score of these actionlets $\text{Conf}_{S'} < \text{Conf}_S$ is less than the confidence threshold.

The outline of the mining process is shown in Algorithm 1. For each class $c$, the mining algorithm outputs a discriminative actionlet pool $P_c$ which contains the actionlets that meet our criteria: $\text{Amb}_S \leq T_{\text{amb}}$ and $\text{Conf}_S \geq T_{\text{conf}}$.

The speed of the proposed data mining algorithm is a lot faster than naively enumerating all the candidate actionlets. We implement the proposed data mining algorithm with Python and run it on a Corei7-2600K machine with 8 GB memory. In our experiment on MSR-DailyActivity3D dataset, which contains 20 human joints and 320 sequences, we set the threshold for confidence score $T_{\text{conf}} = -1$, and the threshold for ambiguity score $T_{\text{amb}} = -2$. The data mining algorithm generates 180 actionlets in 5.23 s. In contrast, naively enumerating all the candidate actionlets takes 307 s under the same environment.

Since we do not impose the constraints that the discriminative actionlets are significantly different from each other, there may be some redundancies among the discovered discriminative actionlets. We will employ multiple kernel learning algorithm to select the discriminative actionlets as described in the next subsection.

### 2.4.2 Learning Actionlet Ensemble

The discriminative power of a single actionlet is limited. In this subsection, we propose to learn an *actionlet ensemble* with multiple kernel learning approach.

| | |
|---|---|
| 1 | Take the set of joints, the feature $G_j$ on each joint $j$, the number of the classes $C$, thresholds $T_{\text{conf}}$ and $T_{\text{amb}}$. |
| 2 | Train the base classifier on the features $G_j$ of each joint $j$. |
| 3 | **for** *Class $c = 1$ to $C$* **do** |
| 4 |     Set $P_c$, the discriminative actionlet pool for class $c$ to be empty : $P_c = \{\}$. Set $l = 1$. |
| 5 |     **repeat** |
| 6 |         Generate the $l$-actionlets by adding one joint into each $(l - 1)$-actionlet in the discriminative actionlet pool $P_c$. |
| 7 |         Add the $l$-actionlets whose confidence scores are larger than $T_{\text{conf}}$ to the pool $P_c$. |
| 8 |         $l = l + 1$ |
| 9 |     **until** *no discriminative actionlet is added to $P_c$ in this iteration*; |
| 10 |     remove the actionlets whose ambiguities scores are larger than $T_{\text{amb}}$ in the pool $P_c$. |
| 11 | **end** |
| 12 | **return** the discriminative actionlet pool for all the classes. |

**Algorithm 1:** Discriminative Actionlet Mining

An *actionlet ensemble* is a linear combination of the actionlet classifiers. For each actionlet $S_k$ in the discriminative actionlet pool, we train an SVM model on it as an actionlet classifier, which defines a joint feature map $\boldsymbol{\Phi}_k(x, y)$ on data $\mathscr{X}$ and labels $\mathscr{Y}$ as a linear output function $f_k(x, y)$ parameterized with the hyperplane normal $w_k$ and bias $b_k$:

$$\begin{aligned} f_k(x, y) &= \langle w_k, \boldsymbol{\Phi}_k(x, y) \rangle + b_k \\ &= \sum_i \alpha_{ik} K_k((x_i, y_i), (x, y)) + b_k \end{aligned} \tag{2.14}$$

where each kernel $K_k(., .)$ corresponds to the conjunctive features of $k$-th actionlet. We employ linear kernel in our actionlet ensemble model. Thus, given two conjunctive features of the $k$-th actionlet $x_i^k$ and $x_j^k$, their kernel function can be represented as $K_k(x_i^k, x_j^k) = (x_i^k)^T x_j^k$. The predicted class $y$ for $x$ is chosen to maximize the output $f_k(x, y)$.

Multiclass-MKL considers a convex combination of $p$ kernels, $K(x_i, x_j) = \sum_{k=1}^p \beta_k K_k(x_i, x_j)$. Equivalently, we consider the following output function:

$$f_{\text{final}}(x, y) = \sum_{k=1}^p [\beta_k \langle w_k, \boldsymbol{\Phi}_k(x, y) \rangle + b_k] \tag{2.15}$$

We aim at choosing $w = (w_k), b = (b_k), \beta = (\beta_k), k = 1, \ldots, p$, such that given any training data pair $(x^{(i)}, y^{(i)})$, $f_{\text{final}}(x^{(i)}, y^{(i)}) \geq f_{\text{final}}(x^{(i)}, u)$ for all $u \in \mathscr{Y} - \{y^{(i)}\}$. The resulting optimization problem becomes:

$$\begin{aligned} \min_{\beta, w, b, \xi} \quad & \frac{1}{2} \boldsymbol{\Omega}(\beta, w) + C \sum_{i=1}^n \xi_i \\ \text{s.t. } \forall i : \, & \xi_i = \max_{u \neq y_i} l(f_{\text{final}}(x^{(i)}, y^{(i)}) - f_{\text{final}}(x^{(i)}, u)) \end{aligned} \tag{2.16}$$

where $C$ is the regularization parameter and $l(f) = \max(0, 1 - f)$ is a convex hinge loss function, and $\Omega(\beta, w)$ is a regularization term on $\beta$ and $w$.

Following the approach in [38], we choose $\Omega(\beta, w) = \|\beta\|_1^2 + C_2\|w\|_2^2$. Since there exists redundancies among the discriminative actionlets discovered with the data mining algorithm, the $l_1$ regularization $\|\beta\|_1^2$ acts as a feature selection regularization by encouraging a sparse $\beta$, so that an ensemble of a small number of non-redundant actionlets is learned. The regularization $\|w\|_2^2$ encourages the actionlet classifiers to have large margin.

This problem can be solved by iteratively optimizing $\beta$ with fixed $w$ and $b$ through sparse solver, and optimizing $w$ and $b$ with fixed $\beta$ through a generic SVM solver such as LIBSVM.

## 2.5 Experimental Results

We choose CMU MoCap dataset [2], MSR-Action3D dataset [3], MSR-DailyActivity 3D dataset, Cornell Activity dataset [4], and Multiview 3D Event dataset to evaluate the proposed action recognition approach. In all the experiments, we use two-level Fourier Temporal Pyramid, with 1/4 length of each segment as low-frequency coefficients. The coefficients of all levels are concatenated sequentially. The empirical results show that the proposed framework outperforms the state-of-the-art methods.

### 2.5.1 MSR-Action3D Dataset

MSR-Action3D dataset [3] is an action dataset of depth sequences captured by a depth camera. This dataset contains twenty actions: *high arm wave, horizontal arm wave, hammer, hand catch, forward punch, high throw, draw x, draw tick, draw circle, hand clap, two hand wave, side-boxing, bend, forward kick, side kick, jogging, tennis swing, tennis serve, golf swing, pick up & throw*. Every action was performed by ten subjects three times each. The frame rate is 15 frames per second and resolution $640 \times 480$. Altogether, the dataset has 402 action sequences with a total of 23,797 frames of depth maps. Some examples of the depth sequences are shown in Fig. 2.2.

Those actions were chosen to cover a variety of movements of arms, legs, torso and their combinations. The subjects were advised to use their right arm or leg if an action is performed by a single arm or leg. Although the background of this dataset is clean, this dataset is challenging because many of the actions in the dataset are highly similar to each other (See Fig. 2.6).

The 3D joint positions are extracted from the depth sequence by using the real time skeleton tracking algorithm proposed in [1]. Since there is no human-object interaction in this dataset, we only extract the 3D joint position features in this experiment(Table 2.1).

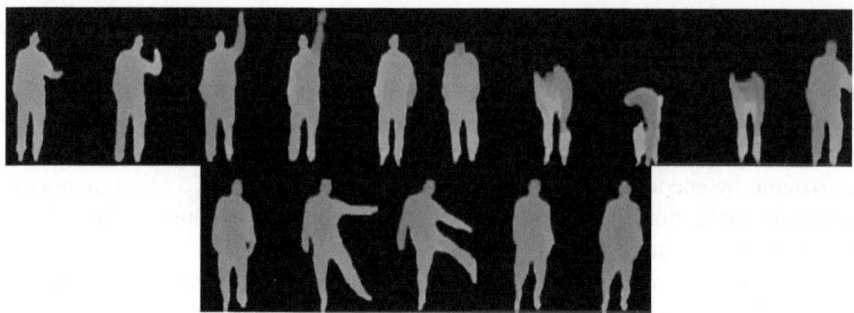

**Fig. 2.6**  Sample frames of the MSR-Action3D dataset

**Table 2.1** Recognition accuracy comparison for MSR-Action3D dataset

| Method | Accuracy |
|---|---|
| Recurrent neural network [14] | 0.425 |
| Dynamic temporal warping [15] | 0.540 |
| Hidden markov model [10] | 0.630 |
| Action graph on bag of 3D points [3] | 0.747 |
| Histogram of 3D joints [39] | 0.789 |
| Random occupancy pattern [25] | 0.862 |
| Eigenjoints [27] | 0.823 |
| Sequence of most informative joints [29] | 0.471 |
| Proposed method with absolute joints positions | 0.685 |
| Proposed method | **0.882** |

We compare our method with the state-of-the-art methods on the cross-subject test setting [3], where the examples of subjects 1, 3, 5, 7, 9 are used as training data, and the examples of subjects 2, 4, 6, 8, 10 are used as testing data. The recognition accuracy of the dynamic temporal warping is only 54 %, because some of actions in the dataset are very similar to each other, and there are typical large temporal misalignment in the dataset. The accuracy of recurrent neural network is 42.5 %, while he accuracy of Hidden Markov Model is 63 %. The recently proposed joint-based action recognition methods, including Histogram of 3D joints, Eigenjoints and Sequence of most informative joints, achieve accuracy 78.9, 82.3 and 47.06 %, respectively. The proposed method achieves an accuracy of 88.2 %. This is a very good performance considering that the skeleton tracker sometimes fails and the tracked joint positions are quite noisy. We also compare the proposed relative joint position features with the absolute joint position features. The proposed method using absolute joint positions achieves much worse accuracy than the proposed method using relative joint positions.

The confusion matrix is illustrated in Fig. 2.7. For most of the actions, our method works very well. The classification errors occur if two actions are too similar to each other, such as "hand catch" and "high throw", or if the occlusion is so large that the skeleton tracker fails frequently, such as the action "pick up and throw"(Fig. 2.7).

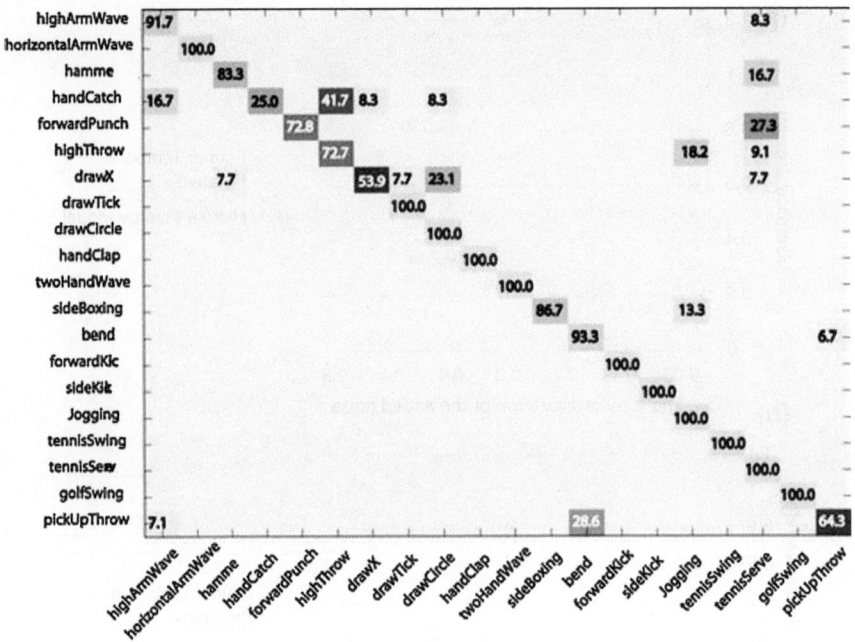

**Fig. 2.7** The confusion matrix for MSR-Action3D dataset

The proposed temporal representation Fourier Temporal Pyramid has two advantages: robustness to the noise and temporal misalignment, which are common in the action sequences captured with Kinect camera. In this experiment, we compare its robustness with a widely utilized temporal representation: Hidden Markov Model. The comparison of the noise robustness of the Fourier Temporal Pyramid features and that of Hidden Markov Model is shown in Fig. 2.8a. In this experiment, we add white Gaussian noise to the 3D joint positions of the samples, and compare the relative accuracies of the two methods. For each method, its relative accuracy is defined as the accuracy under the noisy environment divided by the accuracy under the noiseless environment. We can see that the proposed Fourier Temporal Pyramid feature is much more robust to noise than Hidden Markov Model, because the clustering algorithm employed in Hidden Markov Model to obtain hidden states is relatively sensitive to noise, especially when the different actions are similar to each other.

The temporal shift robustness of the proposed method and the Hidden Markov model is also compared. In this experiment, we circularly shift all the training data, and keep the testing data unchanged. The relative accuracy is shown in Fig. 2.8b. Hidden Markov Model is very robust to the temporal misalignment, because learning a Hidden Markov Model does not require the sequences to be temporally aligned. We find that the proposed approach is also robust to the temporal shift of the depth sequences, though the Fourier Temporal Pyramid is slightly more sensitive to temporal shift.

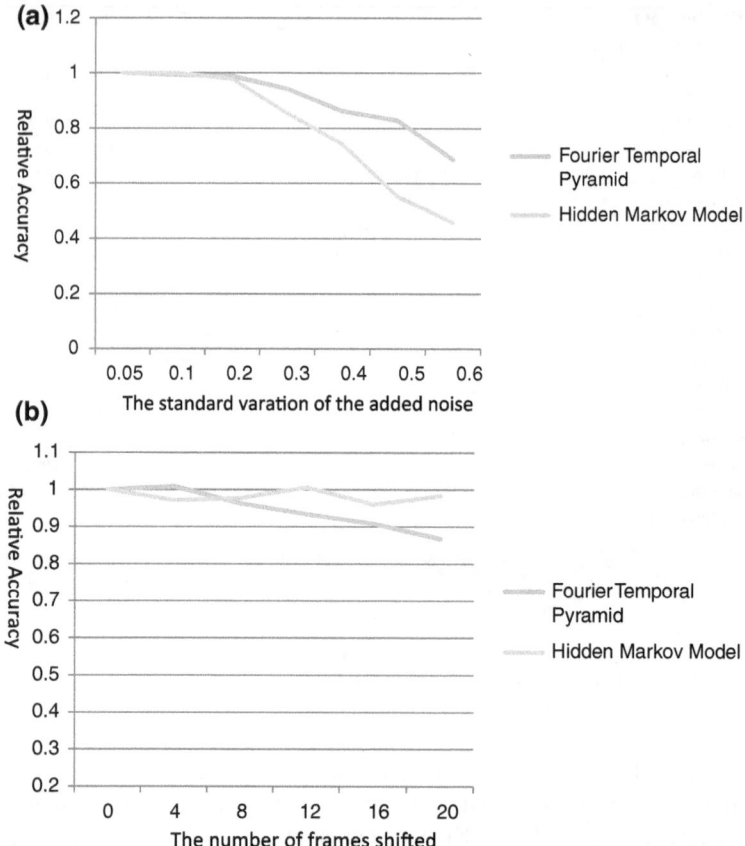

**Fig. 2.8** The relationship between the relative accuracy and the variance of noise or temporal misalignment. **a** The standard variation of the added noise. **b** The number of frames shifted

Thus, compared with widely applied Hidden Markov Model, the proposed Fourier Temporal Pyramid is a temporal representation that exhibits more robustness to noise while retaining the Hidden Markov Model's robustness to temporal misalignment. These properties are important for the action recognition with the depth maps and joint positions captured by Kinect devices, which can be very noisy and contain strong temporal misalignment.

Another advantage of the proposed Fourier Temporal Pyramid is its robustness to the number of action repetitions in the sequences. In order to evaluate the robustness of the proposed method to the number of action repetitions, we manually replicate all the action sequences two times and four times and apply the proposed algorithm to the new sequences. The recognition accuracy is 86.45 and 86.83 % for two-times repetitions and four-times repetitions, respectively. If we repeat half of the action sequences two times, and the other half of the action sequences four times, the

**Fig. 2.9** The recognition accuracy of the proposed Actionlet Ensemble method using different levels of Fourier Pyramid

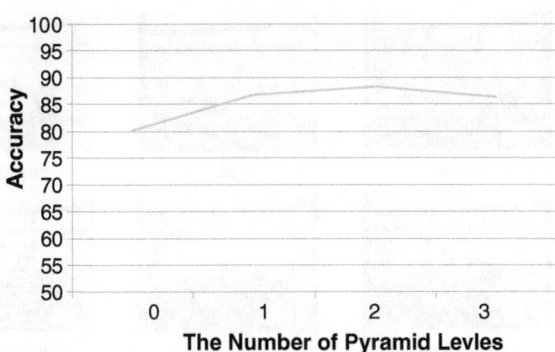

recognition accuracy is 84.24 %. This experiment shows that the proposed method is relatively insensitive to the number of action repetitions.

We also study the relationship between the levels of Fourier pyramid and the recognition accuracy of the proposed Actionlet Ensemble method, and the result is shown in Fig. 2.9. Each level of pyramid divides one temporal segment into two parts. For example, 2-level Fourier Temporal Pyramid contains 1, 2, 4 segments in level 0, 1 and 2, respectively. We can see that the proposed method achieves the best performance when the number of pyramid levels is 2, although the performance is quite close when the number of pyramid levels is 1 or 3.

## 2.5.2  DailyActivity3D Dataset

DailyActivity3D dataset is a daily activity dataset captured by a Kinect device. There are 16 activity types: *drink, eat, read book, call cellphone, write on a paper, use laptop, use vacuum cleaner, cheer up, sit still, toss paper, play game, lay down on sofa, walk, play guitar, stand up, sit down*. If possible, each subject performs an activity in two different poses: "sitting on sofa" and "standing". The total number of the activity sequences is 320. Some example activities are shown in Fig. 2.10.

This dataset is designed to cover daily activities in a living room. This dataset is more challenging than MSR-Action3D dataset. When the performer stands close to the sofa or sits on the sofa, the 3D joint positions extracted by the skeleton tracker are very noisy. Moreover, most of the activities involve the humans-object interactions.

We apply the cross-subject setting to evaluate the proposed algorithm on this dataset. The subjects 1, 2, 3, 4, 5 are used as training data, while the subjects 6, 7, 8, 9, 10 are used as testing data. Table 2.2 shows the accuracies of different methods. By employing an actionlet ensemble model, we obtain a recognition accuracy of 85.75 %. This is a decent result considering the challenging nature of the dataset. If we directly train an SVM on the Fourier Temporal Pyramid features, the accuracy is 78 %. When only the LOP feature is employed, the recognition accuracy

|   Drink   |   Eat   |   Read book   |   write on a paper   |
|  Use laptop   |   play game   |   call cellphone   |   use vacuum cleaner   |
|  Cheer up   |   Sit Still   |   Walking   |   Sit down   |
|  Toss paper   |   lay down on sofa   |   Stand up   |   play guitar   |

**Fig. 2.10**  Sample frames of the DailyActivity3D dataset

**Table 2.2**  Recognition accuracy comparison for MSR-DailyActivity3D dataset

| Method | Accuracy |
| --- | --- |
| Dynamic temporal warping [15] | 0.54 |
| Random occupancy pattern [40] | 0.64 |
| Only LOP features | 0.43 |
| Only joint position features | 0.68 |
| SVM on Fourier Temporal Pyramid features | 0.78 |
| Actionlet ensemble on LOP features | 0.61 |
| Actionlet ensemble on joint features | 0.74 |
| MKL on all the base features | 0.80 |
| Actionlet ensemble | **0.86** |

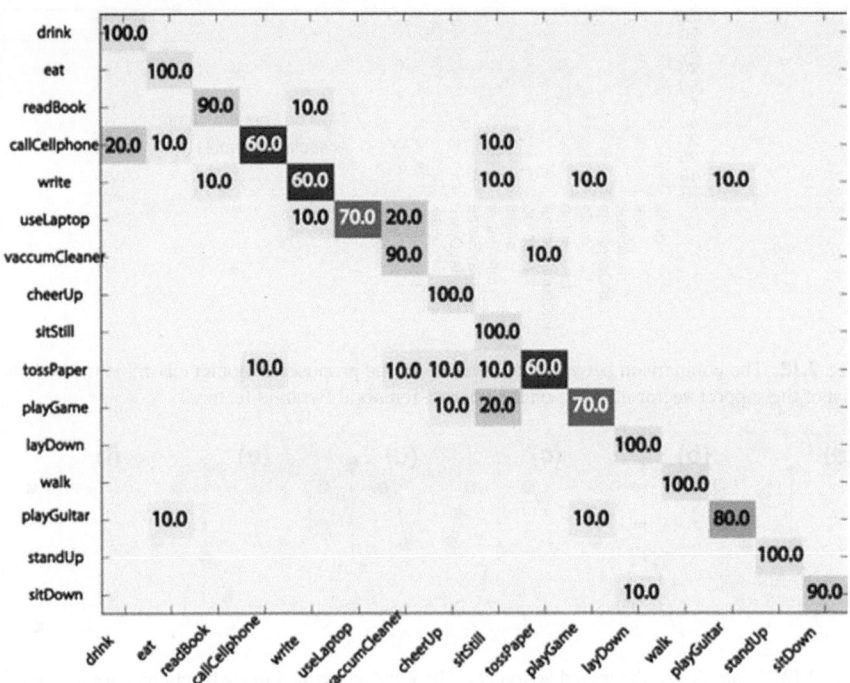

**Fig. 2.11** The confusion matrix of the proposed method on DailyActivity3D dataset

drops to 42.5 %. If we only use 3D joint position features without using LOP, the recognition accuracy is 68 %. If we train an Multiple Kernel Learning classifier on all the base features, the recognition accuracy is 80 %. We also evaluate Random Occupancy Pattern (ROP) [40] method on MSR-DailyActivity3D dataset. Since, the ROP method requires a segmentation of the human, we manually crop out the human, and apply the ROP method on this dataset. The accuracy of ROP method is 64 %.

Figure 2.11 shows the confusion matrix of the proposed method. The proposed approach can successfully discriminate "eating" and "drinking" even though their motions are very similar, because the proposed LOP feature can capture the shape differences of the objects around the hand. Figure 2.12 compares the accuracy of the actionlet ensemble method and that of the support vector machine on the Fourier Temporal Pyramid features. We can observe that for the activities where the hand gets too close to the body, the proposed actionlet ensemble method can significantly improve the accuracy. Figure 2.13 illustrates some of the actionlets with large kernel weights discovered by our mining algorithm.

We also study the effect of the parameters of the actionlet mining algorithm on the recognition accuracy, shown in Fig. 2.14. In this experiment, we adjust $T_{amb}$ while fixing $T_{conf} = -1$ and adjust $T_{conf}$ while fixing $T_{amb} = -1.8$. We find that the proposed data mining algorithm is not sensitive to these two parameters as long as

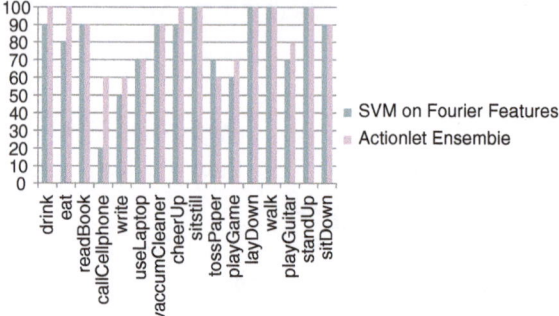

**Fig. 2.12** The comparison between the accuracy of the proposed actionlet ensemble method and that of the support vector machine on the Fourier Temporal Pyramid features

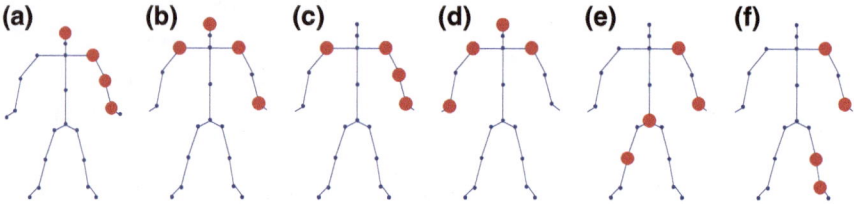

**Fig. 2.13** Examples of the mined actionlets. The joints contained in each actionlet are marked as *red*. **a,b** are actionlets for "drink" **c, d** are actionlets for "call", **e, f** are actionlets for "walk"

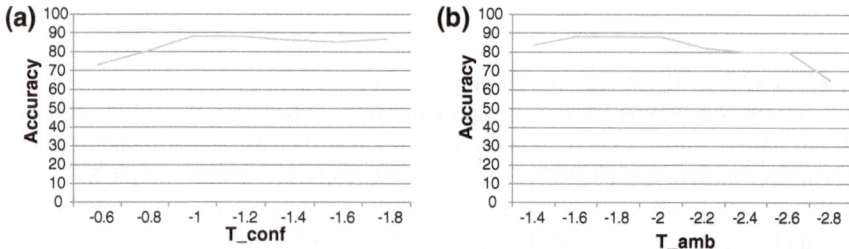

**Fig. 2.14** The relationship between the recognition accuracy and the parameters $T_{conf}$ and $T_{amb}$, which are the threshold of the confidence score and ambiguity score in the actionlet mining algorithm, respectively

they are in a reasonable range. However, setting $T_{conf}$ too high or setting $T_{amb}$ too low may seriously undermine the recognition accuracy because the actionlet mining algorithm rejects discriminative actionlets in these cases. On the other hand, setting $T_{conf}$ too low or setting $T_{amb}$ too high may lead to a large number of the actionlets generated by the actionlet mining algorithm, which greatly slows down the actionlet mining and the actionlet ensemble learning procedures.

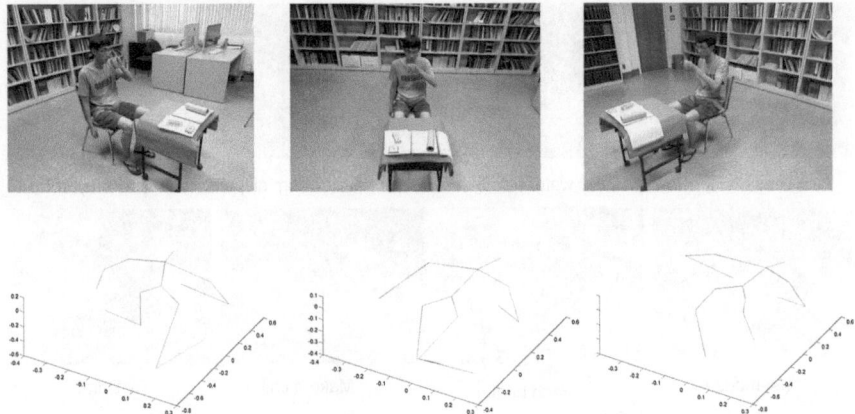

**Fig. 2.15** An action captured from three views and their aligned skeletons

## 2.5.3 Multiview 3D Event Dataset

Multiview 3D event dataset[1] contains RGB, depth and human skeleton data captured simultaneously by three Kinect cameras. This dataset includes 8 event categories: *drink with mug, make a call, read book, use mouse, use keyboard, fetch water from dispenser, pour water from kettle, press button.* Each event is performed by 8 actors. Each actor repeats each event 20 times independently with different object instances and action styles. An example action captured from three view points is illustrated in Fig. 2.15. In total, there are 480 sequences per action class. Figure 2.16 shows some examples of this dataset. The background of this dataset is relatively clean. The difficulty of this dataset is to generalize across different views. We apply our algorithm to this dataset to evaluate the robustness of our algorithm across different views.

Before applying the proposed approach to this dataset, we first perform a human orientation normalization described in Sect. 2.3.4. Although the human body orientations are aligned, action recognition across multiple views is still challenging due to the following two reasons. Firstly, the occlusions of different views are very different even for the same action, as shown in Fig. 2.15. Since the occluded joints are usually non-critical joints ("legs"), these occlusions can be handled effectively by the proposed actionlet ensemble model, because the actionlet ensemble model usually does not contain the non-critical joints. Secondly, the orientation normalization is usually not perfect due to skeleton tracking noise and errors. The proposed actionlet ensemble model is also robust to these noises thanks to the Fourier Temporal Pyramid representation. As a result, the proposed algorithm achieves very good performance on cross-view action recognition experiment.

---

[1] This dataset will be released to public.

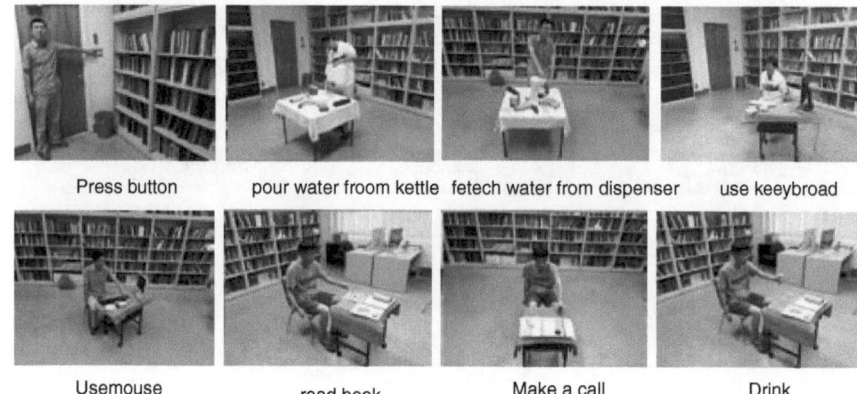

Press button        pour water froom kettle  fetech water from dispenser    use keeybroad

Usemouse            read book              Make a call                  Drink

**Fig. 2.16**  Sample frames of the Multiview 3D event dataset

**Table 2.3**  Recognition accuracy comparison for multiview 3D event dataset

| Method | C-subject | C-view |
|---|---|---|
| Dynamic temporal warping [15] | 0.4712 | 0.4533 |
| Hidden Markov model [10] | 0.848 | 0.6187 |
| Actionlet ensemble | **0.8834** | **0.8676** |

First, we perform cross-subject recognition experiment. In this setting, we use examples of 1/3 of the subjects as training data and the rest of the examples as testing data. We implement the dynamic temporal warping [15] and Hidden Markov Model [10] and compare the proposed model to these models. The classification accuracy comparison of these algorithms is shown in Table 2.3. The dynamic temporal warping achieves 47.12 %, and the hidden Markov model achieves 84.8 % accuracy on this setting, while the proposed algorithm achieves an accuracy of 88.34 %. Most of the confusion occurs between the actions "drinking" and "make a phone call", because the movement of these two actions are very similar.

Then, we perform evaluation under the cross-view recognition setting. In this setting, the data are partitioned into three subsets each corresponding to a different camera. We use the subset from one camera as the testing data and use the data from the other two cameras for training. Three-fold cross-validation is applied to measure the overall accuracy. The results are listed in Table 2.3. One observation is that the proposed algorithm is quite robust across multiple views. The proposed algorithm achieves an accuracy of 86.75 % on cross-view setting, which is only 1.4 % lower than the accuracy on cross-subject setting. The confusion matrix is shown in Fig. 2.18.

The experimental results show that, with orientation normalization, the proposed algorithm can achieve good action recognition accuracy under cross-view action recognition setting.

**Fig. 2.17** The confusion matrix for multiview 3D event dataset on cross-subject setting

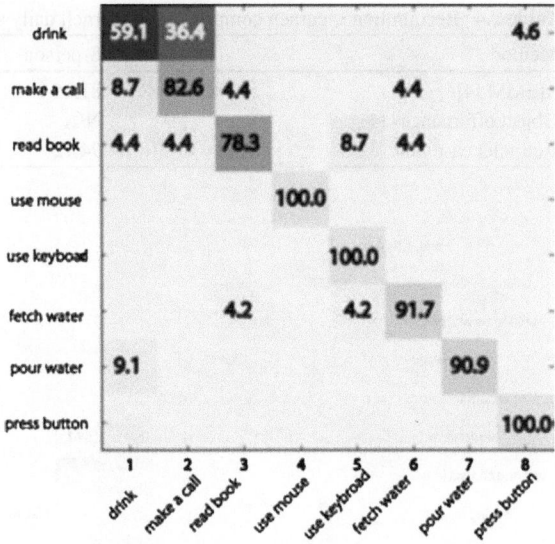

**Fig. 2.18** The confusion matrix for multiview 3D event dataset on cross-view setting

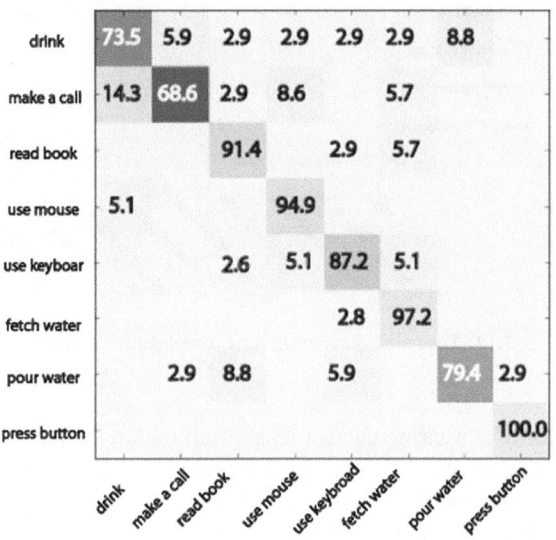

## 2.5.4  Cornell Activity Dataset

Cornell Activity dataset (CAD-60) [4] contains the RGB frames, depth sequences and the tracked skeleton joint positions captured with Kinect cameras. The actions in this dataset can be categorized into 5 different environments: office, kitchen, bedroom, bathroom, and living room. Three or four common activities were identified for each environment, giving a total of twelve unique actions: "rinsing mouth", "brushing

**Table 2.4**  Recognition accuracy comparison for cornell daily activity dataset

| Method | S-person | C-person |
|---|---|---|
| MEMM [4] | 81.15 | 51.9 |
| Object offordances [41] | N/A | 71.4 |
| Actionlet ensemble | **94.12** | **74.70** |

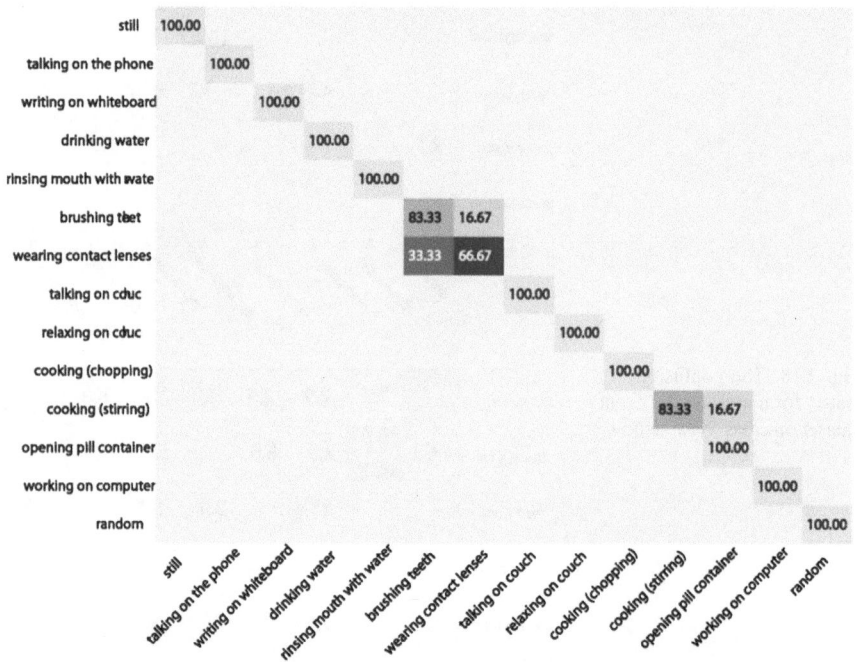

**Fig. 2.19**  The confusion matrix for Cornell Activity dataset on same-person setting

teeth", "wearing contact lens", "talking on the phone", "drinking water", "opening pill container", "cooking (chopping)", "cooking (stirring)", "drinking water", "talking on couch", "relaxing on couch", "talking on the phone", "writing on whiteboard", "drinking water", "working on computer"

The recognition accuracy is shown in Table 2.4. We employ the same experimental setup as [4]: The same-person experiment setup employs half of the data of the same person as training, and the other half is used as testing. The cross-person experiment setup uses leave-one-person-out cross-validation. The proposed method achieves an accuracy of 97.06 % for the same-person setup and 74.70 % for the cross-person setup. Both results are better than those of the state-of-the-art methods.

The confusion matrices of the proposed algorithm on Cornell Activity dataset under the same-person setting and the cross-subject setting are shown in Figs. 2.19 and 2.20, respectively. We can see that the proposed algorithm correctly classifies

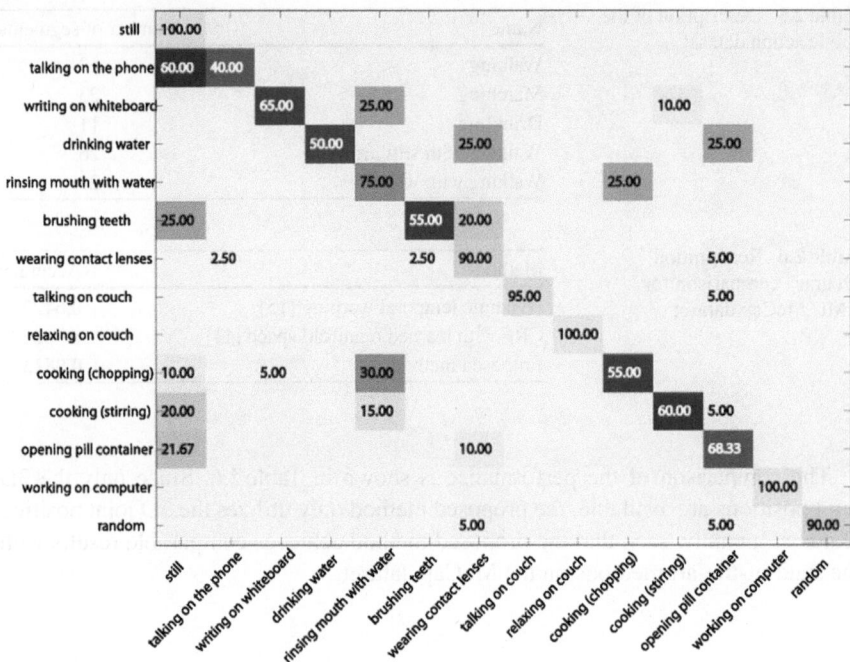

**Fig. 2.20**   The confusion matrix for Cornell Activity dataset on cross-person setting

most of the actions under the same-person setting. The cross-person setting is more challenging, and we find that many actions are classified into "still" under this setting, because the motions of these actions are very subtle and there are serious noises in Kinect skeleton tracking. Thus, it is difficult to distinguish the action "still" from those actions with subtle motions.

## 2.5.5  CMU MoCap Dataset

We also evaluate the proposed method on the 3D joint positions extracted by a motion capture system. The dataset we use is the CMU Motion Capture (MoCap) dataset.

Five subtle actions are chosen from CMU MoCap datasets following the configuration in [11]. The five actions differ from each other only in the motion of one or two limbs. The actions in this dataset include: *walking, marching, dribbling, walking with stiff arms, walking with wild legs*. The number of segments of each action is listed in Table 2.5. The 3D joint positions in CMU MoCap dataset are relatively clean because they are captured with high-precision camera array and markers. This dataset is employed to evaluate the performance of the proposed 3D joint position-based features on 3D joint positions captured by Motion Capture system.

**Table 2.5** Description of the subtle action dataset

| Name | Number of segments |
|------|--------------------|
| Walking | 69 |
| Marching | 23 |
| Dribbling | 11 |
| Walking with stiff arms | 26 |
| Walking with wild legs | 28 |

**Table 2.6** Recognition accuracy comparison for CMU MoCap dataset

| Method | Accuracy |
|--------|----------|
| Dynamic temporal warping [15] | 0.6427 |
| CRF with learned manifold space [11] | 0.9827 |
| Proposed method | **0.9813** |

The comparison of the performance is shown in Table 2.6. Since only the 3D joint positions are available, the proposed method only utilizes the 3D joint position features. It can be seen that the proposed method achieves comparable results with the state-of-the-art methods on the MoCap dataset.

## 2.6 Conclusion

In this chapter, we propose a novel actionlet ensemble model that characterizes the conjunctive structure of 3D human actions by capturing the correlations of the joints that are representative of an action class. We also propose two novel features to represent 3D human actions with depth and skeleton data. Local occupancy pattern describes "depth appearance" in the neighborhood of a 3D joint. Fourier Temporal Pyramid describes the temporal structure of an action. The proposed features effectively discriminate human actions with subtle differences and human-object interactions and are robust to noise and temporal misalignment. Our extensive experiments demonstrated the superior performance of the proposed approach to the state-of-the-art methods. In the future, we aim to exploit the effectiveness of the proposed technique for the understanding of more complex activities.

## References

1. Shotton, J., Fitzgibbon, A., Cook, M., Sharp, T., Finocchio, M., Moore, R., Kipman, A., Blake, A.: Real-time human pose recognition in parts from single depth images. In: CVPR (2011)
2. "CMU graphics lab motion capture database", http://www.mocap.cs.cmu.edu/
3. Li, W., Zhang, Z., Liu, Z.: Action recognition based on a bag of 3d points. In: Human Communicative Behavior Analysis Workshop (in Conjunction with CVPR) (2010)
4. Sung, J., Ponce, C., Selman, B., Saxena, A.: Unstructured human activity detection from RGBD images. In: International Conference on Robotics and Automation (2012)

5. Wang, J., Liu, Z., Wu, Y., Yuan, J.: Mining actionlet ensemble for action recognition with depth cameras. In: CVPR (2012)
6. Laptev, I.: On space-time interest points. IJCV **64**(2–3), 107–123 (2005)
7. Laptev, I., Marszalek, M., Schmid, C., Rozenfeld, B.: Learning realistic human actions from movies. In: CVPR, pp. 1–8 (2008)
8. Dalal, N., Triggs, B., Histograms of oriented gradients for human detection. In: IEEE CVPR, pp. 886–893 (2005)
9. Campbell, L.W., Bobick, A.F.: Recognition of human body motion using phase space constraints. In: ICCV (1995)
10. Lv, F., Nevatia, R.: Recognition and segmentation of 3-D human action using HMM and Multi-class AdaBoost. In: ECCV, pp. 359–372 (2006)
11. Han, L., Wu, X., Liang, W., Hou, G., Jia, Y.: Discriminative human action recognition in the learned hierarchical manifold space. Image Vis. Comput. **28**(5), 836–849 (2010)
12. Ning, H., Xu, W., Gong, Y., Huang, T.: Latent pose estimator for continuous action. In: ECCV, pp. 419–433 (2008)
13. Chen, H.S., Chen, H.T., Chen, Y.W., Lee, S.Y.: Human action recognition using star skeleton. In: Proceedings of the 4th ACM International Workshop on Video Surveillance and Sensor Networks, pp. 171–178, New York, USA (2006)
14. Martens J., Sutskever, I.: Learning recurrent neural networks with hessian-free optimization. In: ICML (2011)
15. Muller, M., Röder, T.: Motion templates for automatic classification and retrieval of motion capture data. In: Proceedings of the 2006 ACM SIGGRAPH/Eurographics symposium on Computer animation, pp. 137–146, Eurographics Association (2006)
16. Li L., Prakash, B.A. Time series clustering: complex is simpler! In: ICML (2011)
17. Dai, S., Yang, M., Wu, Y., Katsaggelos, A.: Detector ensemble. In: IEEE CVPR, pp. 1–8 (2007)
18. Zhu, L., Chen, Y., Lu, Y., Lin, C., Yuille, A.: Max margin AND/OR graph learning for parsing the human body. In: IEEE CVPR (2008)
19. Yuan, J., Yang, M., Wu, Y.: Mining discriminative co-occurrence patterns for visual recognition. In: CVPR (2011)
20. Yao, B., Fei-Fei, L.: Grouplet: a structured image representation for recognizing human and object interactions. In: CVPR (2010)
21. Agrawal, R., Srikant, R.: Fast algorithms for mining association rules. In: VLDB, vol. 1215, pp. 487–499 (1994)
22. Bourdev L., Malik, J.: Poselets: body part detectors trained using 3d human pose annotations. In: CVPR (2009)
23. Desai C., Ramanan, D.: Detecting actions, poses, and objects with relational phraselets. In: ECCV (2012)
24. Maji, S., Bourdev, L., Malik, J.: Action recognition from a distributed representation of pose and appearance. In: IEEE CVPR (2011)
25. Wang, J., Liu, Z., Chorowski, J., Chen, Z., Wu, Y.: Robust 3D action recognition with random occupancy patterns. In: ECCV, pp. 1–14 (2012)
26. Vieira, A.W., Nascimento, E.R.. Oliveira, G.L., Liu, Z., Campos, M.M.: STOP: space-time occupancy patterns for 3D action recognition from depth map sequences. In: 17th Iberoamerican Congress on Pattern Recognition, Buenos Aires (2012)
27. Yang X., Tian, Y.: EigenJoints-based action recognition using Naïve-Bayes-Nearest-Neighbor. In: CVPR 2012 HAU3D, Workshop (2012)
28. Yang, X., Zhang, C., Tian, Y.L.: Recognizing actions using depth motion maps-based histograms of oriented gradients. In: ACM Multimedia (2012)
29. Ofli, F., Chaudhry, R., Kurillo, G., Vidal, R., Bajcsy, R.: Sequence of the most informative joints (smij): a new representation for human skeletal action recognition. J. Visual Commun Image Represent. **26**, 1140–1145 (2013)
30. Yun, K., Honorio, J., Chattopadhyay, D., Berg, T.L. Samaras, D., Brook, S.: Two-person interaction detection using body-pose features and multiple instance learning. In: CVPR 2012 HAU3D Workshop (2012)

31. Raptis, M., Kirovski, D., Hoppe, H.: Real-time classification of dance gestures from skeleton animation. In: Proceedings of the 2011 ACM SIGGRAPH/Eurographics Symposium on Computer Animation—SCA '11, p. 147. ACM Press, New York, NY, USA (2011)
32. Chaudhry, R., Ofli, F., Kurillo, G., Bajcsy, R., and Vidal, R.: Bio-inspired dynamic 3D discriminative skeletal features for human action recognition. In: HAU3D13 (2013)
33. Lazebnik, S., Schmid, C., Ponce, J.: Beyond bags of features: spatial pyramid matching for recognizing natural scene categories. In: IEEE CVPR, vol. 2 (2006)
34. Oppenheim, A.V., Schafer, R.W., Buck, J.R.: Discrete Time Signal Processing (Prentice Hall Signal Processing Series). Prentice Hall, Upper Saddle River (1999)
35. Fischler M.A., Bolles, R.C.: Random sample consensus: a paradigm for model fitting with applications to image analysis and automated cartography. Commun. ACM **24**(6), 381–395 (1981)
36. Chapelle, O., Vapnik, V., Bousquet, O., Mukherjee, S.: Choosing multiple parameters for support vector machines. Mach. Learn. **46**(1), 131–159 (2002)
37. Wu, T.-F., Lin, C.-J. Weng, R.C.: Probability estimates for multi-class classification by pairwise coupling. JMLR **5**, 975–1005 (2004)
38. Friedman, J.H., Popescu, B.E.: Predictive learning via rule ensembles. Ann. Appl. Stat. **2**(3), 916–954 (2008)
39. Xia, L., Chen, C.-C., Aggarwal, J.K.: View invariant human action recognition using histograms of 3D joints The University of Texas at Austin. In: CVPR 2012 HAU3D Workshop (2012)
40. Wang, J., Yuan, J., Chen, Z., Wu, Y.: Spatial locality-aware sparse coding and dictionary learning. In: ACML (2012)
41. Koppula, H.S., Gupta, R., Saxena, A.: Learning human activities and object affordances from RGB-D videos. arXiv preprint arXiv:1210.1207 (2012)

# Chapter 3
# Random Occupancy Patterns

**Abstract** The introduction of the depth cameras has opened up a new way for spatio-temporal pattern classification by providing the depth map of a scene, but the unique characteristics of the depth maps also calls for novel spatio-temporal representations. The depth maps do not have as much texture as the conventional RGB images do, and they are much more noisy. When the depth maps are captured from just a single view, occlusion is another serious problem. In order to deal with these issues, we develop a semi-local feature called random occupancy pattern (ROP), which employs a novel progressive rejection sampling scheme to effectively explore an extremely large sampling space. We also utilize a sparse coding approach to robustly encode these features. The proposed approach does not require careful parameter tuning. Its training is very fast due to the use of the high-dimensional integral image and the efficient sampling scheme, and it is robust to the occlusions. Our technique is evaluated on three datasets captured by commodity depth cameras: an action dataset and a hand gesture dataset. Our classification results are comparable or superior to those obtained by the state-of-the-art approaches on all two datasets. The experiments also demonstrate the robustness of the proposed method to noises and occlusions.

**Keywords** Random occupancy pattern · Occlusion · Sparse coding · Extreme learning machine

## 3.1 Introduction

Recently, the advance of the imaging technology has enabled us to capture the depth information in real-time, and various promising applications have been proposed [1–4]. Compared with conventional cameras, the depth camera has several advantages. For example, segmentation in depth images is much easier, and depth images are insensitive to changes in lighting conditions.

J. Wang et al., *Human Action Recognition with Depth Cameras*,
SpringerBriefs in Computer Science, DOI: 10.1007/978-3-319-04561-0_3,
© The Author(s) 2014

Although skeleton tracking algorithm proposed in [1] is very robust when little occlusion occurs, it can produce inaccurate results or even fails when serious occlusion occurs. Moreover, the skeleton tracking is unavailable for human hands thus cannot be utilized for hand gesture recognition.

Therefore, we aim at developing an action recognition approach that directly takes the depth sequences as input. Designing an efficient depth sequences representation for action recognition is a challenging task. First of all, depth sequences may be seriously contaminated by occlusions, which makes the global features unstable. On the other hand, the depth maps do not have as much texture as color images do, and they are too noisy to apply local differential operators such as gradients on. These challenges motivate us to seek for features that are semi-local, highly discriminative and robust to occlusion.

In this chapter, we treat a three-dimensional action sequence as a 4D shape and propose random occupancy pattern (ROP) features, which are extracted from randomly sampled 4D subvolumes with different sizes and at different locations. Since the ROP features are extracted at a larger scale, it is robust to noise. At the same time, they are less sensitive to occlusion because they only encode information from the regions that are most discriminative for the given action.

An Elastic-Net regularized classification model is developed to further select the most discriminative features, and sparse coding is utilized to robustly encode the features. The feature encoding step further improves the proposed method's robustness to occlusion by modeling the occlusion noise as the sparse coding reconstruction errors. The proposed approach performs well on the depth sequence dataset and is robust to occlusion. The general framework of the proposed method is shown in Fig. 3.1.

We evaluate our method on two datasets captured by commodity depth cameras. The experimental results validate the effectiveness of the proposed method.

Our main contributions are as follows: First, we propose a computationally efficient method to perform action recognition from depth sequences. Second, a novel weighted sampling scheme is proposed to effectively explore an extremely large dense sampling space. Third, we propose to employ sparse coding to deal with the occlusions in the depth sequences.

## 3.2 Related Work

The Haar wavelet-like features have been successfully applied in [5] for face detection. A boosted classifier is learned by applying AdaBoost algorithm [6] on a very large pool of weak classifiers. The weak classifier pool is constructed based on the features extracted from the rectangles at all possible locations and scales. While this approach is successful for 2D images, when the data becomes 3D or 4D, the number of possible rectangles becomes so large that enumerating them or performing AdaBoost algorithm on them becomes computationally prohibitive. Weinland et al. [7] utilizes 3D occupancy features and models the dynamics by an exemplar-based

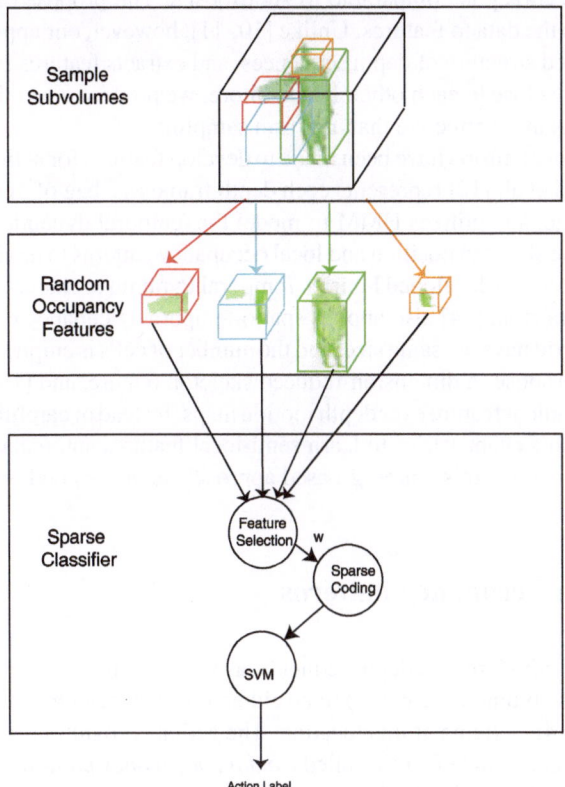

**Fig. 3.1** The framework of the proposed method. The 3D subvolumes are shown for illustration purpose. In the implementation, 4D subvolumes are employed

hidden Markov model. The proposed ROP feature is much simpler and more computationally efficient than Haar-like features, while achieving similar performances in depth datasets.

Randomization has been applied in [8, 9] to address this problem. Bangpeng et al. [9] employs a random forest to learn discriminative features that are extracted either from a patch or from a pair of patches for fine-grained image categorization. Amit and Geman [8] apply a random forest to mine discriminative features from binary spatial arrangement features for character recognition. Their work demonstrates the effectiveness of randomization in dealing with large sampling space. We exploit randomization to perform action recognition from depth sequences, in which the data is sparse and the number of the possible subvolumes is much larger.

Rahimi and Recht [10] and Huang et al. [11] also employ randomization in the learning process. These approaches randomly map the data to features with a linear function whose weights and biases are uniformly sampled. Their empirical and theoretical results show that their training is much faster than AdaBoost's, while their classification accuracy is comparable to AdaBoost's. The proposed approach also randomly maps the data to features. Unlike [10, 11], however, our approach exploits the neighborhood structure of depth sequences, and extracts features from the pixels that are spatially close to each other. Furthermore, we propose a weighted sampling technique that is more effective than uniform sampling.

Recently, a lot of efforts have been made to develop features for action recognition in depth data. Li et al. [12] represents each depth frame as a bag of 3D points on the human silhouette, and utilizes HMM to model the temporal dynamics. Wang et al. [13] uses relative skeleton position and local occupancy patterns to model the human-object interaction, and developed Fourier Temporal Pyramid to characterize temporal dynamics. Vieira et al. [14] also applies spatio-temporal occupancy patterns, but all the cells in the grid have the same size, and the number of cells is empirically set. Yang and Tian [15] proposes a dimension-reduced skeleton feature, and [16] developed a histogram of gradient feature over depth motion maps. Instead of carefully developing good features, this chapter tries to learn semi-local features automatically from the data, and we show that this learning-based approach achieves good results.

## 3.3  Random Occupancy Patterns

The proposed method treats a depth sequence as a 4D volume, and defines the value of a pixel in this volume $I(x, y, z, t)$ to be either 1 or 0, depending on whether there is a point in the 4D volume at this location. The action recognition is performed by using the value of a simple feature called *random occupancy patterns*, which is both efficient to compute and highly discriminative.

We employ four dimensional random occupancy patterns to construct the features, whose values are defined to be the soft-thresholded sum of the pixels in a subvolume:

$$o_{xyzt} = \delta(\sum_{q \in \text{bin}_{xyzt}} I_q) \tag{3.1}$$

where $I_q = 1$ if the point cloud has a point in the location $q$ and $I_q = 0$ otherwise. $\delta(.)$ is a sigmoid normalization function: $\delta(x) = \frac{1}{1+e^{-\beta x}}$. This feature is able to capture the occupancy pattern of a 4D subvolume. Moreover, it can be computed in constant complexity with the high dimensional integral images [17].

As shown in Fig. 3.1, we extract ROP features from the subvolumes with different sizes and at different locations. However, the number of possible simple features is so large that we are not able to enumerate all of them. In this chapter, we propose a weighted random sampling scheme to address this problem, which will be described in Sect. 3.4.

## 3.4 Weighted Sampling Approach

Since the number of possible positions and sizes of the 4D subvolumes is extremely large and the information of these features is highly redundant, it is neither necessary nor computationally efficient to explore all of them. In this section, we propose a random sampling approach to efficiently explore these 4D subvolumes.

### 3.4.1 Dense Sampling Space

Recall that a ROP feature is extracted from a 4D subvolume. It is computationally prohibitive to extract the features from all possible subvolumes, thus we would like to sample a subset of discriminative subvolumes and extract the features from them. This subsection characterizes the *dense sampling space*, from which we randomly sample subvolumes.

Denote the size of a depth sequence volume to be $W_x \times W_y \times W_z \times W_t$. A subvolume can be characterized by two points $[x_0, y_0, z_0, t_0]$ and $[x_1, y_1, z_1, t_1]$, and is denoted as $[x_0, y_0, z_0, t_0] \sim [x_1, y_1, z_1, t_1]$. A normal subvolume has the property that $x_0 \leq x_1, y_0 \leq y_1, z_0 \leq z_1$, and $t_0 \leq t_1$, and the subvolume is the set of points

$$\{[x, y, z, t] : x_0 \leq x \leq x_1, y_0 \leq y \leq y_1,$$
$$z_0 \leq z \leq z_1, t_0 \leq t \leq t_1\} \qquad (3.2)$$

Our sampling space consists of all the subvolumes $[x_0, y_0, z_0, t_0] \sim [x_1, y_1, z_1, t_1]$ where $x_0, x_1 \in \{1, 2, \cdots, W_x\}$, $y_0, y_1 \in \{1, 2, \cdots, W_y\}$, $z_0, z_1 \in \{1, 2, \cdots, W_z\}$, $t_0, t_1 \in \{1, 2, \cdots, W_t\}$. If we take $(W_x, W_y, W_z, W_t) = (80, 80, 80, 80)$, the size of the dense sampling space is $80^8 = 1.67 \times 10^{15}$. This dense sampling space is so large that exhaustively exploring it is computationally prohibitive. However, since the subvolumes highly overlap with each other, they contain redundant information, and it is possible to employ randomization to deal with this problem.

### 3.4.2 Weighted sampling

One way to sample from the dense sampling space is to perform uniform sampling. Nevertheless, since in the depth sequences many subvolumes do not contain useful information for classification, uniform sampling is highly inefficient. In this section, to efficiently sample from the dense sampling space, we propose a weighted sampling approach based on the rejection sampling, which samples the discriminative subvolumes with high probability.

To characterize how discriminative a subvolume is, we employ the scatter matrix class separability measure [18]. The scatter matrices include *Within-class scatter matrix* ($S_W$), *Between-class scatter matrix* ($S_B$), and *Total scatter matrix* ($S_T$). They are defined as $S_W = \sum_{i=1}^{c} \sum_{j=1}^{n_i} (h_{i,j} - m_i)(h_{i,j} - m_i)^T$, $S_B = \sum_{i=1}^{c} n_i (m_i - m)(m_i - m)^T$, and $S_T = S_W + S_B$, where $c$ is the number of the classes, $n_i$ denotes the number of training data in the $i$-th class, and $h_{i,j}$ denotes the features extracted from the $j$-th training data in the $i$-th class. $m_i$ denotes the mean vector of the features $h_{i,j}$ in the $i$-th class and $m$ the mean vector of the features extracted from all the training data. A large separability measure means that these classes have small within-class scatter and large between-class scatter, and the class separability measure $J$ can be defined as

$$J = \frac{\text{tr}(S_W)}{\text{tr}\, S_B} \tag{3.3}$$

Denote $V$ as the 4D volume of a depth sequence. For each pixel $p \in V$, we define a neighborhood subvolume centered at $p$, and extract the eight Haar feature values from this neighborhood subvolume. These eight feature values form an 8-dimensional vector which is used as the feature vector $h$ to evaluate the class separability score $J_p$ at pixel $p$.

A subvolume should be discriminative if all the pixels in this subvolume are discriminative, and vice versa. Therefore, we utilize the average of the separability scores of all the pixels in the region $R$ as the separability score of $R$.

The probability that a subvolume $R$ is sampled should be proportional to its separability score $J_R$, that is,

$$P_{R\,\text{sampled}} \propto J_R = \frac{1}{N_R} \sum_{p \in R} J_p \tag{3.4}$$

where $N_R$ is the number of pixels in the subvolume $R$.

We can uniformly draw a subvolume, and accept the subvolume with probability

$$P_{R\,\text{accept}} = \frac{W_x W_y W_z W_t}{\sum_{p \in V} J_p} J_R \tag{3.5}$$

Note that $P_{R\,\text{uniform}} P_{R\,\text{accept}} = P_{R\,\text{sampled}}$. Therefore, with the rejection sampling scheme, the probability that $R$ is selected is equal to the desired probability as specified in Eq. (3.4). The derivation of the acceptance rate can be found in the supplemental material.

Because we are able to compute the average separability score in a subvolume very efficiently using the high dimensional integral image, this sampling scheme is computationally efficient. The outline of the algorithm is shown in Algorithm 1.

---

1  Let $N_S$ denote the number of subvolumes to sample, $J_p$ the separability score at pixel $p$.

2  Compute $P = \dfrac{W_x W_y W_z W_t}{\sum_{p \in V} J_p}$

3  **repeat**

4      Uniformly draw a subvolume $R$.

5      Compute the average separability score in this subvolume $J_R$ with integral image.

6      Compute the acceptance rate $P_{accept} = P J_R$.

7      Uniformly draw a number $N$ from $[0, 1]$.

8      **if** $N \leq P_{accept}$ **then**

9          Retain the subvolume $R$.

10      **end**

11      **else**

12          Discard the subvolume $R$.

13      **end**

14  **until** *the number of subvolumes sampled* $\geq N_S$;

---

**Algorithm 1**: Weighted Sampling Algorithm

## 3.5 Learning Classification Functions

Given the training data pairs $(x^i, t^i)$, $i = 1, \cdots, n$, where $x^i \in \mathscr{R}^L$ denotes a training data, and $t^i \in \mathscr{T}$ is the corresponding label, the aim of the classification problem is to learn a prediction function $g : \mathscr{R}^L \to \mathscr{T}$. Without loss of generality, we assume the classification problem is a binary classification problem, i.e., $\mathscr{T} = \{0, 1\}$. If the problem is a multiclass problem, it can be converted into a binary classification problem with the one-vs-others approach.

An Elastic-Net regularization is employed to select a sparse subset of features that are the most discriminative for the classification. Choosing a sparse subset of features has several advantages. First, the speed of the final classifier is faster if the number of the selected features is smaller. Second, learning a sparse classification function is less prone to over-fitting if only limited amount of training data is available [19].

For each training data sample $x^i$, $N_f$ ROP features are extracted: $h^i_j$, $j = 1, \cdots, N_f$, and the response is predicted by a linear function

$$y^i = \sum_{j=1}^{N_f} w_j h^i_j \tag{3.6}$$

Denote $w$ as the vector containing all $w_j$, $j = 1, \cdots, N_f$. The objective of the learning is to find $w$ that minimizes the following objective function:

$$E = \sum_{i=1}^{n} (t^i - y^i)^2 + \lambda_1 \|w\|_1 + \lambda_2 \|w\|_2^2 \tag{3.7}$$

where $\lambda_1$ is a regularization parameter to control the sparsity of $w$, and $\lambda_2$ is used to ensure that the margin of the classifier is large. It has been shown that if the number of features $N_f$ is much larger than that of the training data $n$, which is the case of this chapter, Elastic-Net regularization works particularly well [19]. SPAMS toolbox [20] is employed to numerically solve this optimization problem.

The selected feature $f$ is obtained by discarding the features $x_j$ with corresponding $w_j$ less than a given threshold and multiplying the rest of $x_j$ by weight $w_j$.

All the training data is utilized as the dictionary $A = [f^1, f^2, \cdots, f^n]$. For a test data with feature $f$, we can solve the following sparse coding problem:

$$\min \frac{1}{2}\|f - A\alpha\|_2^2 + \lambda\|\alpha\|_1 \tag{3.8}$$

where $\alpha$ is called the reconstruction coefficients, and $\lambda$ is a regularization parameter. This model assumes the feature vector $f$ can be represented as the linear combination of the features of the training data plus a noise vector $\varepsilon$.

$$f = \sum_{i=1}^{n} \alpha_i f^i + \varepsilon \tag{3.9}$$

The reconstruction coefficients $\alpha$ are employed to represent a depth sequence, and an SVM classifier is trained for action classification.

## 3.6 Experimental Results

In this section, we evaluate our algorithm on two datasets captured by commodity depth cameras: the MSR-Action3D datasets [12] and Gesture3D dataset. The experimental results show that our algorithm outperforms the existing methods on these datasets, and is not sensitive to occlusion error. The $\beta$ of the sigmoid function for ROP feature is set to be 10 in all the experiments.

### 3.6.1 MSR-Action3D

MSR-Action3D dataset [12] is an action dataset of depth sequences captured by a depth camera. This dataset contains 20 actions: *high arm wave, horizontal arm wave, hammer, hand catch, forward punch, high throw, draw x, draw tick, draw circle, hand clap, two hand wave, side-boxing, bend, forward kick, side kick, jogging, tennis swing, tennis serve, golf swing, pick up and throw*. Each action was performed by 10 subjects for three times. The frame rate is 15 frames per second and the resolution is $640 \times 480$. Altogether, the dataset has 23,797 frames of depth maps for 402 action samples. Some examples of the depth map sequences are shown in Fig. 3.2.

Those actions were chosen to cover various movements of arms, legs, torso and their combinations, and the subjects were advised to use their right arm or leg if an

**Fig. 3.2** Sample frames of the MSR-Action3D dataset

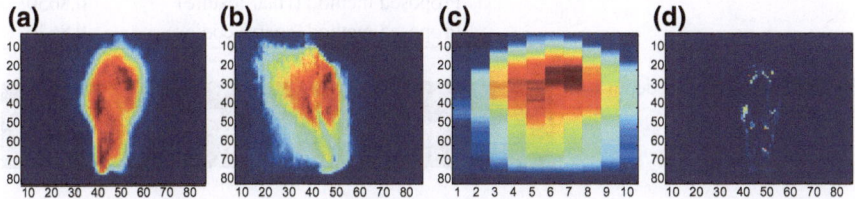

**Fig. 3.3** The projection of the separability scores for MSR-Action3D, where *red color* means high intensity. **a** The projection of the separability scores to x-y plane. **b** The projection of the separability scores to y-z plane. **c** The projection of the separability scores to y-t plane. **d** The cross section of the separability scores on x-y plane at $z = 40, t = 1$

action is performed by a single arm or leg. Although the background of this dataset is clean, this dataset is challenging because many of the actions in the dataset are highly similar to each other. In this experiment, 50,000 subvolumes are sampled unless otherwise stated.

In order to ensure the consistency of the scale, each depth sequence is resized to the same size $80 \times 80 \times 80 \times 10$. The separability scores of the MSR-Action3D are shown in Fig. 3.3. The regions with high separability scores are human's arms and legs, which is consistent with the characteristics of the dataset, and the beginning and the ending parts of an action are less discriminative than the middle part of an action. Moreover, in Fig. 3.3d, it can be observed that the center of the human does not have high separability score, because the center of the human usually does not have many movements and does not contain useful information for classification.

We compare our algorithm with the state-of-the-art method [12] on this dataset, which extracts the contour information from the depth maps, where the examples of subjects $1, 3, 5, 7, 9$ are used as training data, and the examples of subjects $2, 4, 6, 8, 10$ are used as testing data. Table 3.1 shows the recognition accuracy. The recognition accuracy is computed by running the experiments 10 times and taking the average of each experiment's accuracy. Our method outperforms this method by a large margin. Notice that our classification configuration uses half of the subjects as the training data and the rest of them as test data, which is difficult because of the larger variations across the same actions performed by different subjects. Our method is also compared with the STIP features, Laptev [21], which is a state-of-the-art local feature designed for action recognition from RGB videos. The local spatio-temporal features do not work well for depth data because there is little texture in depth maps. Another method we compare with is the convolutional network.

**Table 3.1** Recognition
accuracy comparison for
MSR-Action3D dataset

| Method | Accuracy |
|---|---|
| STIP features [21] | 0.423 |
| Action graph on bag of 3D points [12] | 0.747 |
| High dimensional convolutional network | 0.725 |
| STOP feature [14] | 0.848 |
| Eigenjoints [15] | 0.823 |
| Support vector machine on raw data | 0.79 |
| Proposed method (without sparse coding) | 0.8592 |
| Proposed method (Haar feature) | 0.8650 |
| Proposed method (sparse coding) | 0.8620 |

**Fig. 3.4** An occluded depth
sequence

**Table 3.2** Robustness to occlusion comparison

| Occlusion | Accuracy without using sparse coding | Accuracy using sparse coding |
|---|---|---|
| 1 | 83.047 | 86.165 |
| 2 | 84.18 | 86.5 |
| 3 | 78.76 | 80.09 |
| 4 | 82.12 | 85.49 |
| 5 | 84.48 | 87.51 |
| 6 | 82.46 | 87.51 |
| 7 | 80.10 | 83.80 |
| 8 | 85.83 | 86.83 |

We have implemented a 4-dimensional convolutional network by extending the three-dimensional convolutional network of [22]. Finally we compare with a Support Vector Machine classifier on the raw features consisting of the pixels on all the locations. Although the Support Vector Machine performs surprisingly well on our dataset, the training time of the SVM is very long because the dimension of the features is very high. In contrast, the proposed method is simple to implement and is computationally efficient in both training and testing. Moreover, it outperforms all the other methods including SVM. In addition, we can see that the proposed ROP feature performs comparably with Haar features.

In order to test the sensitivity of the proposed method to occlusions, we divide each depth sequences into $2 \times 2 \times 1 \times 2$ subvolumes, i.e., we partition each depth sequences into two parts in y, x and t dimensions. Each volume only covers half of the frames of the depth sequences. Occlusion is simulated by ignoring the points that fall into the specified occluded subvolume, illustrated in Fig. 3.4. We run the simulation with one subvolume occluded, the performance is shown in Table 3.2. It can seen that employing sparse coding can greatly improve the robustness.

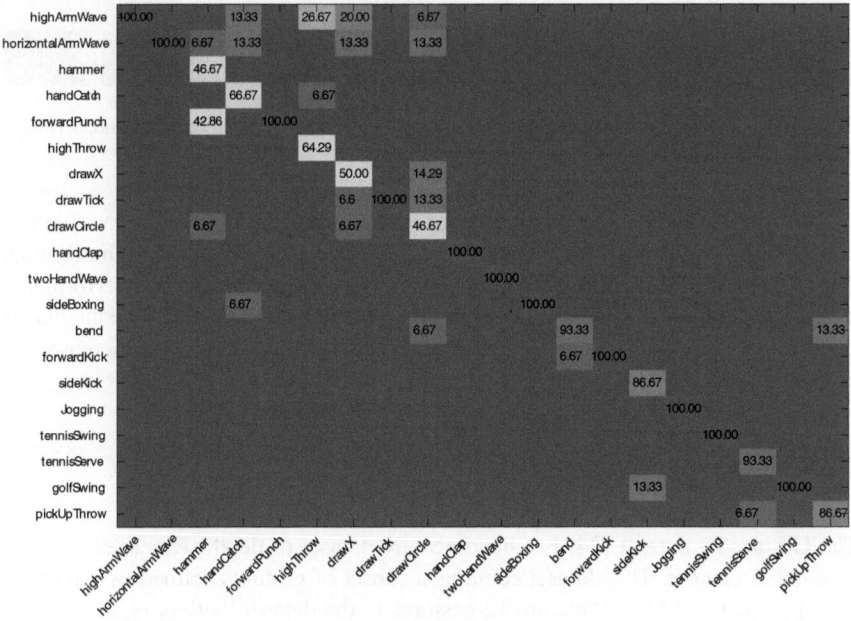

**Fig. 3.5** The confusion matrix for the proposed method on MSR-Action3D dataset. It is recommended to view the figure on the screen

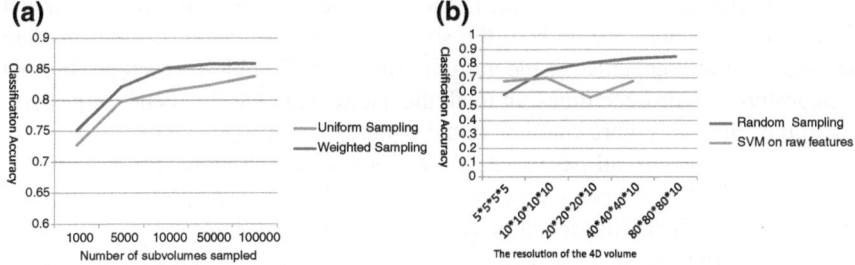

**Fig. 3.6** The comparison between different sampling methods, and the relationship between the resolution of data and the classification accuracy for SVM and the proposed sampling method

The confusion matrix is shown in Fig. 3.5. The proposed method performs very well on most of the actions. Some actions, such as "catch" and "throw", are too similar to each other for the proposed method to capture the difference.

We also compare the classification accuracy of the proposed sampling scheme to that of the uniform sampling scheme in Fig. 3.6a. It can be observed that the weighted sampling scheme is more effective than the uniform sampling scheme. Moreover, the proposed scheme does not suffer from overfitting even when the number of the sampled subvolumes is very large. Bartlett [23] gives an intuitive proof of the generalization ability of classifier of the randomly generated features.

**(a)**                              **(b)**                              **(c)**

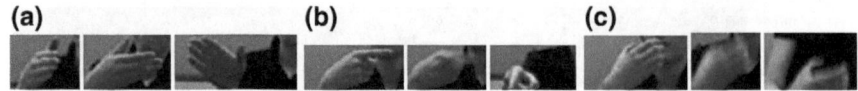

**Fig. 3.7**   The sample frames of the Gesture3D dataset, **a** the blue gesture. **b** Green gesture. **c** Hungry gesture

The depth sequences are downsampled into different resolutions, and we explore the relationship between the resolution of the data and the classification accuracy. The relationship found is shown in Fig. 3.6b. Our observation is that the performance of the SVM classifier may drop when we increase the resolution of the data, but for our random sampling scheme, increasing the data resolution always increases the classification accuracy.

### 3.6.2 Gesture3D Dataset

The Gesture3D dataset [24] is a hand gesture dataset of depth sequences captured by a depth camera. This dataset contains a subset of gestures defined by American Sign Language (ASL). There are 12 gestures in the dataset: *bathroom, blue, finish, green, hungry, milk, past, pig, store, where, j, z*. Some example frames of the gestures are shown in Fig. 3.7. Notice that although this dataset contains both the color and depth frames, only depth frames are used in the experiments. Further description of the gestures can be found in [25]. All of the gestures used in this experiment are dynamic gestures, where both the shape and the movement of the hands are important for the semantics of the gesture. There are 10 subjects, each performing each gesture two or three times. In total, the dataset contains 336 depth sequences. The self occlusion is more common in the gesture dataset.

In this experiment, all gesture depth sequences are subsampled to size $120 \times 120 \times 3 \times 10$. The leave-one-subject-out cross-validation is employed to evaluate the proposed method. The recognition accuracy is shown in Table 3.3. The proposed method performs significantly better than the SVM on raw features and the high dimensional convolutional network. Our performance is also slight better than the action graph model which uses carefully designed shape features [24].

The separability score map for Gesture3D dataset is shown in Fig. 3.8. We observe that the score map of the gestures is very different from that of the actions shown in Fig. 3.3, because the movement pattern of the gestures and the actions is very different. In gesture depth sequences, the semantics of the gesture are mainly determined by the large movement of the hand, while the human actions are mainly characterized by the small movements of the limbs.

The confusion matrix is shown in Fig. 3.9. The proposed method performs quite well for most of the gestures. It can be observed from the confusion matrix that large confusion exists between the gesture "where" and "green". Both gestures involve the movement of one finger, and only the directions of the movement are slightly different.

**Fig. 3.8** The projection of the scores for Gesture3D, **a** the projection of the scores to x-y plane. **b** The projection of the scores to y-t plane

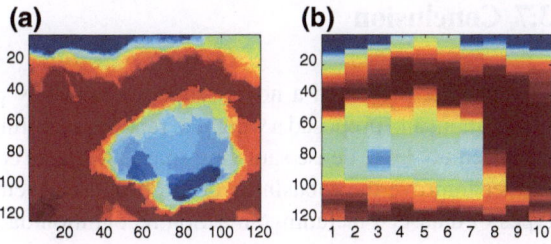

**Table 3.3** Recognition accuracy comparison for Gesture3D dataset

| Method | Accuracy |
|---|---|
| SVM on raw features | 0.6277 |
| High dimensional convolutional network [22] | 0.69 |
| Action graph on occupancy features [24] | 0.805 |
| Action graph on Silhouette features [24] | 0.877 |
| Proposed method (without sparse coding) | 0.868 |
| Proposed method (sparse coding) | 0.885 |

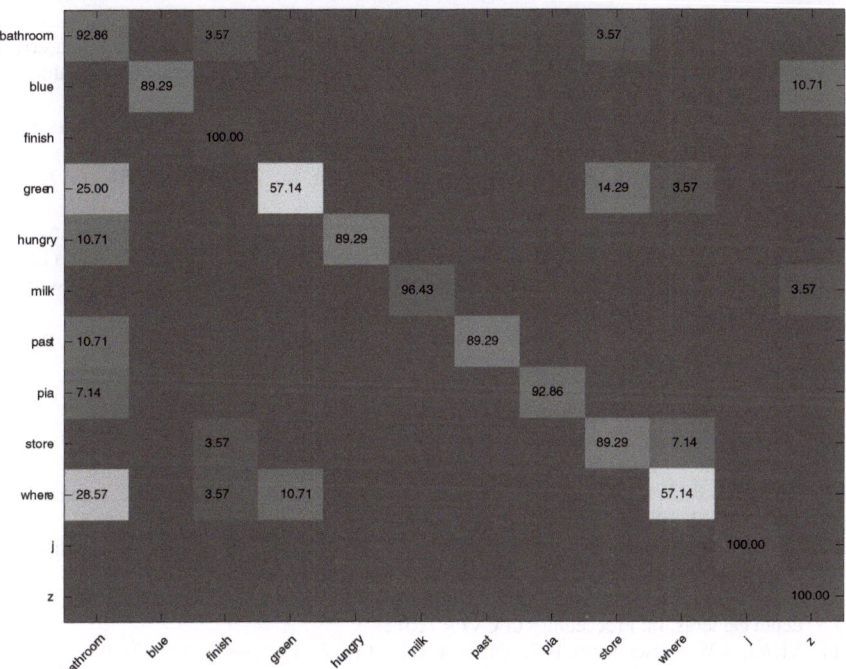

**Fig. 3.9** The confusion matrix of the proposed method on Gesture3D dataset

## 3.7 Conclusion

This chapter presented a novel random occupancy pattern features for 3D action recognition, and proposed a weighted random sampling scheme to efficiently explore an extremely large dense sampling space. A sparse coding approach is employed to further improve the robustness of the proposed method. Experiments on different types of datasets, including an action recognition dataset and a gesture recognition dataset, demonstrated the effectiveness and robustness of the proposed approach as well as its broad applicability.

## References

1. Shotton, J., Fitzgibbon, A., Cook, M., Sharp, T., Finocchio, M., Moore, R., Kipman, A., Blake, A.: Real-time human pose recognition in parts from single depth images. In: Proceedings of CVPR (2011)
2. Hadfield, S., Bowden, R.: Kinecting the dots :particle based scene flow from depth sensors. In: Proceedings of ICCV (2011)
3. Baak, A., Meinard, M., Bharaj, G., Seidel, H., Theobalt, C., Informatik, M.P.I.: A data-driven approach for real-time full body pose reconstruction from a depth camera. In: Proceedings of ICCV (2011)
4. Girshick, R., Shotton, J., Kohli, P., Criminisi, A., Fitzgibbon, A.: Efficient regression of general-activity human poses from depth images. In: Proceedings of ICCV (2011)
5. Viola, P., Jones, M.J.: Robust real-time face detection. Int. J. Comp. Vis. **57**(2), 137–154 (2004)
6. Freund, Y., Schapire, R.: A decision-theoretic generalization of on-line learning and an application to boosting. In: Computational Learning Theory, vol. 55, pp. 23–37. Springer (1995)
7. Weinland, D., Boyer, E., Ronfard, R.: Action recognition from arbitrary views using 3D exemplars. In: Proceedings of ICCV, pp. 1–7, Oct 2007
8. Amit, Y., Geman, D.: Shape quantization and recognition with randomized trees. Neural Comput. **9**(7), 1545–1588 (1997)
9. Yao, B., Khosla, A., Fei-Fei, L.: Combining randomization and discrimination for fine-grained image categorization. In: Proceedings of CVPR (2011)
10. Rahimi, A., Recht, B.: Weighted sums of random kitchen sinks: replacing minimization with randomization in learning. In: NIPS, vol. 885. Citeseer (2008)
11. Huang, G.-B., Wang, D.H., Lan, Y.: Extreme learning machines: a survey. Int. J. Mach. Learn. Cybern. **2**(2), 107–122 (2011)
12. Li, W., Zhang, Z., Liu, Z.: Action recognition based on a bag of 3d points. In: Human Communicative Behavior Analysis Workshop (in conjunction with CVPR) (2010)
13. Wang, J., Liu, Z., Wu, Y., Yuan, J.: Mining actionlet ensemble for action recognition with depth cameras. In: Proceedings of CVPR (2012)
14. Vieira, A.W., Nascimento, E.R., Oliveira, G.L., Liu, Z., Campos, M.M.: STOP: space-time occupancy patterns for 3D action recognition from depth map sequences. In: 17th Iberoamerican Congress on Pattern Recognition. Buenos Aires (2012)
15. Yang, X., Tian, Y.: EigenJoints-based action recognition using naïve-bayes-nearest-neighbor. In: CVPR 2012 HAU3D, Workshop (2012)
16. Yang, X., Zhang, C., Tian, Y.: Recognizing actions using depth motion maps-based histograms of oriented gradients. In: ACM Multimedia (2012)
17. Tapia, E.: A note on the computation of high-dimensional integral images. Pattern Recogn. Lett. **32**(2), 197–201 (2011)

18. Wang, L., Chan, K.L.: Learning kernel parameters by using class separability measure. In: Proceedings of NIPS (2002)
19. Zou, H., Hastie, T.: Regularization and variable selection via the elastic net. J. Roy. Stat. Soc. **67**(2), 301–320 (2005)
20. Mairal, J.: SPArse modeling software. http://www.di.ens.fr/willow/SPAMS/"
21. Laptev, I.: On space-time interest points. Int. J. Comput. Vis. **64**(2–3), 107–123 (2005)
22. Ji, S., Xu, W., Yang, M., Yu, K.: 3D convolutional neural networks for human action recognition. In: Proceedings of ICML. Citeseer (2010)
23. Bartlett, P.L.: The sample complexity of pattern classification with neural networks: the size of the weights is more important than the size of the network. IEEE Trans. Inf. Theory **44**(2), 525–536 (1998)
24. Kurakin, A., Zhang, Z., Liu, Z.: A real-time system for dynamic hand gesture recognition with a depth sensor. In: Proceedings of EUSIPCO (2012)
25. Basic America Sign Language. http://www.lifeprint.com/asl101/pages-layout/concepts.htm

# Chapter 4
# Conclusion

**Abstract** This chapter concludes the methods that we introduce in this book, and presents the current challenges and further directions in 3D action recognition using depth cameras.

**Keywords** Occlusion · Human-object interaction · Cross-domain recognition

## 4.1 Conclusion

We present two effective action recognition schemes using depth camera. The Actionlet Ensemble method utilizes human skeleton positions tracked by Kinect camera for human movements, and depth maps for human/object interactions. It employs the Fourier Temporal Pyramid as a robust representation for the temporal structures of human actions, and constructs a compositional structure of mined discriminative actionlets. The Random Occupancy Pattern features do not rely on skeletons. Thus, it can not only be used for action recognition, but can also be employed for hand gesture and object recognition. It achieves robustness to occlusions through selecting the most discriminative spatio-temporal patterns by random sampling and discriminative sparse selection.

The action recognition systems have achieved great successes in recent years. It has been used in commercial systems, such as Kinect games and traffic monitoring systems. The performance of the state-of-the-art action recognition algorithms already performs very well on benchmark datasets. However, a lot of challenges still need to be solved to build practical human action recognition systems.

First, handling realistic occlusions is still challenging. The benchmark dataset usually contains no occlusion or synthetic occlusions. In practical applications, the subject can be completely occluded for a few frames, and strong self-occlusion can occur. Accurate action recognition under these occlusions is still challenging.

J. Wang et al., *Human Action Recognition with Depth Cameras*,
SpringerBriefs in Computer Science, DOI: 10.1007/978-3-319-04561-0_4,
© The Author(s) 2014

Second, real-time action recognition requires us to simultaneously segment and classify actions in a video. The real world actions do not have clear separation boundaries over time. It is difficult to perform action recognition and action segmentation independently. Joint action segmentation and recognition methods are required.

Third, there are a lot of rooms for improvement in human-object interaction modeling. The human-object interactions can have complex semantics. The interactions are determined by the type of objects, how the objects are used, and the order in which the interactions are performed. Even the subtle difference in the order of the interactions can lead to completely different semantics. We need more sophisticated semantic models, such as stochastic grammar [1], to model complex human-object interactions.

Finally there is no guarantee that all action recognition systems have depth cameras. A cross-domain systems can relax the requirement of deploying a depth camera while exploiting the depth map information. For example, it may be possible to develop a system that learns a model offline from both RGB videos and depth sequences while performing action recognition on RGB videos without depth data. This type of system can be easily deployed to the environments where deploying depth camera is not feasible.

# Reference

1. Si, Z., Zhu, S.-C.: Learning AND-OR templates for object recognition and detection. PAMI **35**(9), 2189–2205 (2013)

# Index

J. Wang et al., *Human Action Recognition with Depth Cameras*,
SpringerBriefs in Computer Science, DOI: 10.1007/978-3-319-04561-0,
© The Author(s) 2014